AF902891

Springer Theses

Recognizing Outstanding Ph.D. Research

Aims and Scope

The series "Springer Theses" brings together a selection of the very best Ph.D. theses from around the world and across the physical sciences. Nominated and endorsed by two recognized specialists, each published volume has been selected for its scientific excellence and the high impact of its contents for the pertinent field of research. For greater accessibility to non-specialists, the published versions include an extended introduction, as well as a foreword by the student's supervisor explaining the special relevance of the work for the field. As a whole, the series will provide a valuable resource both for newcomers to the research fields described, and for other scientists seeking detailed background information on special questions. Finally, it provides an accredited documentation of the valuable contributions made by today's younger generation of scientists.

Theses are accepted into the series by invited nomination only and must fulfill all of the following criteria

- They must be written in good English.
- The topic should fall within the confines of Chemistry, Physics, Earth Sciences, Engineering and related interdisciplinary fields such as Materials, Nanoscience, Chemical Engineering, Complex Systems and Biophysics.
- The work reported in the thesis must represent a significant scientific advance.
- If the thesis includes previously published material, permission to reproduce this must be gained from the respective copyright holder.
- They must have been examined and passed during the 12 months prior to nomination.
- Each thesis should include a foreword by the supervisor outlining the significance of its content.
- The theses should have a clearly defined structure including an introduction accessible to scientists not expert in that particular field.

More information about this series at http://www.springer.com/series/8790

Kaori Fuyuto

Electroweak Baryogenesis and Its Phenomenology

Doctoral Thesis accepted by
Nagoya University, Nagoya, Japan

Springer

Author
Dr. Kaori Fuyuto
Amherst Center for Fundamental Interactions
Department of Physics
University of Massachusetts Amherst
Amherst, US

Supervisor
Prof. Junji Hisano
Department of Physics
Nagoya University
Nagoya, Japan

ISSN 2190-5053 ISSN 2190-5061 (electronic)
Springer Theses
ISBN 978-981-13-1007-2 ISBN 978-981-13-1008-9 (eBook)
https://doi.org/10.1007/978-981-13-1008-9

Library of Congress Control Number: 2018957564

© Springer Nature Singapore Pte Ltd. 2018
This work is subject to copyright. All rights are reserved by the Publisher, whether the whole or part of the material is concerned, specifically the rights of translation, reprinting, reuse of illustrations, recitation, broadcasting, reproduction on microfilms or in any other physical way, and transmission or information storage and retrieval, electronic adaptation, computer software, or by similar or dissimilar methodology now known or hereafter developed.
The use of general descriptive names, registered names, trademarks, service marks, etc. in this publication does not imply, even in the absence of a specific statement, that such names are exempt from the relevant protective laws and regulations and therefore free for general use.
The publisher, the authors and the editors are safe to assume that the advice and information in this book are believed to be true and accurate at the date of publication. Neither the publisher nor the authors or the editors give a warranty, express or implied, with respect to the material contained herein or for any errors or omissions that may have been made. The publisher remains neutral with regard to jurisdictional claims in published maps and institutional affiliations.

This Springer imprint is published by the registered company Springer Nature Singapore Pte Ltd.
The registered company address is: 152 Beach Road, #21-01/04 Gateway East, Singapore 189721, Singapore

Supervisor's Foreword

Baryon asymmetry is one mystery in our Universe. The Standard Model describes most of the particle phenomena very well; however, it cannot explain the asymmetry. So far, various solutions have been invented. Among them is electroweak baryogenesis, in which the asymmetry is produced during electroweak phase transition. Its typical energy scale is $O(100)$ GeV; therefore, the scenario has high testability at particle physics experiments. The discovered Higgs boson with 125 GeV mass indicated that the Standard Model is not able to cause the first-order phase transition necessary for the successful scenario. Moreover, it turned out that CP violation in the Standard Model does not work enough for creating the present asymmetry. These two aspects motivate us to study the next candidate, namely, physics beyond the Standard Model.

In this thesis, Kaori Fuyuto has investigated electroweak baryogenesis in an extended Standard Model. The model newly introduces an extra Higgs doublet scalar, a singlet scalar, and electroweak-interacting fermions. This setup applies to several extensions of the minimal supersymmetric Standard Model. The first-order phase transition is induced by the singlet scalar, and its analysis is done including estimations of sphaleron energy. Interactions between the Higgs doublet and new fermions bring a significant CP-violating process for the asymmetry. The final baryon asymmetry is estimated by the closed time path formalism, which can deal with quantum non-equilibrium phenomenon under the CP violation. These comprehensive analyses have clarified successful parameter regions in explaining the asymmetry. This thesis has also discussed the testability by two measurements: the Higgs coupling and electric dipole moments of electron, neutron, and proton. Taking into account their current and future prospective sensitivities, it has been discussed how the measurements play a complementary role in the verifications.

Nagoya, Japan
August 2018

Prof. Junji Hisano

Parts of this thesis have been published in the following journal articles:

[1] Kaori Fuyuto, and Eibun Senaha, "Improved sphaleron decoupling condition and the Higgs coupling constants in the real singlet-extended standard model," Physical Review D **90**, 015015 (2014), [https://doi.org/10.1103/PhysRevD.90.015015].

[2] Kaori Fuyuto, Junji Hisano and Eibun Senaha, "Toward verification of electroweak baryogenesis by electric dipole moments," Physics Letter B **755** (2016) 491–497, [https://doi.org/10.1016/j.physletb.2016.02.053].

Acknowledgments

First of all, I am grateful to my adviser Junji Hisano for giving me opportunity to study fruitful areas in the particle physics and for useful advice and discussions. I would also appreciate Eibun Senaha for instructive suggestion and for his continuous encouragement and support. Most part of this thesis is based on the discussions and collaboration with them.

I thank all (ex-)members of the E-ken in Nagoya University, especially Hidetoshi Kawase, Natsumi Nagata, and Kazuhiro Tobe, for encouraging me throughout the master's and doctoral course. Further, I wish to thank Shin'ichi Nojiri for his encouragement.

I also thank Wei-Shu Hou and Masaya Kohda for profitable collaboration.

I am thankful to cosmology group in Saga University, especially Koichi Funakubo, for professional discussions.

Finally, I thank my family and friends for their encouragements and supports.

Contents

Chapter 1
Introduction

Abstract Particle physics is based on quantum field theory. It predicts that all particles have a partner with the same mass but opposite charge, which is called antiparticles. The presence was first confirmed by Carl Anderson in 1932, since then, the antiparticles have been established in various experiments. Today, they can be produced in a laboratory and used for a further quest of the Universe.

It would be natural to imagine that both the particle and antiparticle equally exist in the Universe. However, looking around our Universe, everything is made of the particle. Although the antiparticles like μ^+ can be observed in the atmosphere they are just secondary particles produced in association with collisions of cosmic rays. No one knows why only the particle is left in the current Universe, which is one of the mysteries our Universe holds. In this chapter, we give a brief introduction to it and explain necessary conditions for creating the asymmetry.

Keywords Baryon asymmetry of the Universe · Sakharov's criteria

1.1 Baryon Asymmetry of the Universe

When the Universe is at high temperature, both pair production and annihilation of a particle and antiparticle frequently take place. Typical energy of an involved photon is given by the temperature; therefore it comes to lower with decrease of the temperature. Once the photons do not possess enough energy to cause the pair production, only the pair annihilation proceeds. Simultaneously, the rate of the annihilation grows smaller with the expansion of the Universe, and finally, the process freezes.

In this case, however, it is known that the final asymmetry is too small to explain the present Universe that predominantly holds the particle. Therefore, in order to create the current Universe, difference between the number of particle and antiparticle must exist by the era that the pair annihilation ends. The baryon asymmetry of the Universe (BAU) is one of outstanding problems in the particle physics, and its value is indicated by Planck satellite [1]

© Springer Nature Singapore Pte Ltd. 2018

K. Fuyuto, *Electroweak Baryogenesis and Its Phenomenology*, Springer Theses,
https://doi.org/10.1007/978-981-13-1008-9_1

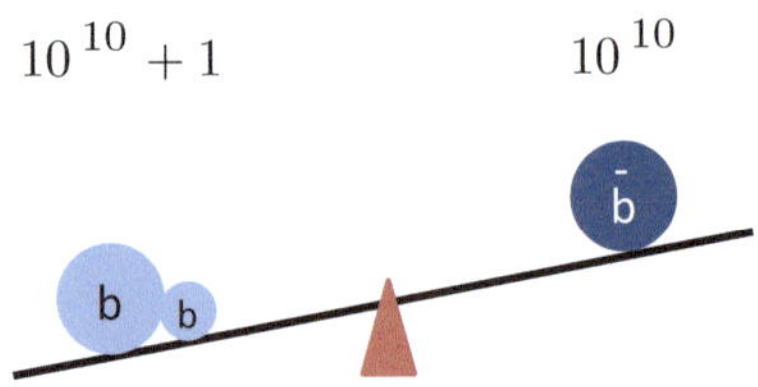

Fig. 1.1 Current cosmological observations show that the number density of baryons (b) is different from that of anti-baryons ($\bar{b}$) by very small amount: $10^{10}+1$ baryons and 10^{10} anti-baryons

$$\frac{n_B}{s} = (8.59 \pm 0.11) \times 10^{-11}, \tag{1.1}$$

where $n_B \equiv n_b - n_{\bar{b}}$ with the baryon (anti-baryon) number density $n_{b(\bar{b})}$ and entropy density s. Unless the baryon number and/or entropy are newly produced, the ratio is invariant under the adiabatic expansion of the Universe. The ratio implies that just a little difference between the baryon and anti-baryon is needed for explanation of the present Universe, whose difference is roughly one to a billion as in Fig. 1.1. It is a surprising fact that the tiny little difference built the Universe; in addition, the situation cannot be explained by the Standard Model (SM). It is called baryogenesis to study how the asymmetry is produced in terms of particle physics.

Before we move onto detailed discussion, let us give a brief comment on two simple but failed stories for the explanation of the BAU. One is to consider that the baryon asymmetry is just an initial condition, namely, the BAU has been present since the birth of the Universe. In this case, unfortunately, exponential expansion of the Universe known as inflation, which is the natural solution of the flatness and horizon problems, would lead to the disappearance of the asymmetry. To be more precise, since the total entropy of the Universe can be produced during the re-heating era after inflation, the initial asymmetry eventually vanishes. Another idea is that the matter exists only around us, but faraway galaxies are being made of antimatter. It implies that the matter and antimatter are separated in the Universe. As discussed in Ref. [2], if galaxies and anti-galaxies come close and collide, the strong gamma rays can be emitted. However, such strong gamma rays have not been seen, so far. As discussed in [3, 4], let us consider the situation where the matter and antimatter are successfully separated before 38 MeV at which $n_b/s = n_{\bar{b}}/s = 8 \times 10^{-10}$. In such a circumstance, the total energy in the horizon amounts to $10^{-7}\,M_\odot$ where $M_\odot$ is solar mass. Then, it turned out that this value cannot explain the present total quantity of matter inside the Milky Way galaxy ($\sim 10^{12}\,M_\odot$).

As mentioned earlier, explanation of the BAU needs some extensions of the SM. So far, various stories for baryogenesis have been invented [5]:

1. GUT baryogenesis. 2. GUT baryogenesis after preheating. 3. Baryogenesis from primordial black holes. 4. String scale baryogenesis. 5. Affleck-Dine (AD) baryogenesis. 6. Hybridized AD baryogenesis. 7. No-scale AD baryogenesis. 8. Single field baryogenesis. 9. Electroweak (EW) baryogenesis. 10. Local EW baryogenesis. 11. Non-local EW baryogenesis. 12. EW baryogenesis at preheating. 13. SUSY EW baryogenesis.

14. String mediated EW baryogenesis. 15. Baryogenesis via leptogenesis. 16. Inflationary baryogenesis. 17. Resonant leptogenesis. 18. Spontaneous baryogenesis. 19. Coherent baryogenesis. 20. Gravitational baryogenesis. 21. Defect mediated baryogenesis. 22. Baryogenesis from long cosmic strings. 23. Baryogenesis from short cosmic strings. 24. Baryogenesis from collapsing loops. 25. Baryogenesis through collapse of vortons. 26. Baryogenesis through axion domain walls. 27. Baryogenesis through QCD domain walls. 28. Baryogenesis through unstable domain walls. 29. Baryogenesis from classical force. 30. Baryogenesis from electrogenesis. 31. B-ball baryogenesis. 32. Baryogenesis from CPT breaking. 33. Baryogenesis through quantum gravity. 34. Baryogenesis via neutrino oscillations. 35. Monopole baryogenesis. 36. Axino induced baryogenesis. 37. Gravitino induced baryogenesis. 38. Radion induced baryogenesis. 39. Baryogenesis in large extra dimensions. 40. Baryogenesis by brane collision. 41. Baryogenesis via density fluctuations. 42. Baryogenesis from hadronic jets. 43. Thermal leptogenesis. 44. Nonthermal leptogenesis.

Here, it should be mentioned that these mechanisms must satisfy three conditions called the Sakharov criteria. Let us see the conditions in next section.

1.2 Sakharov's Criteria

The Sakharov criteria are composed of the following three conditions [6]:

1. Baryon number violating process.
2. C and CP violations.
3. Out of equilibrium.

Here, C and P imply charge and parity symmetries.

As expected easily, the first condition is obviously necessary to create baryon number. One example for the condition is shown in Fig. 1.2, which describes that a particle X decays into particles of Y and b. Here, while the particles of X and Y do not possess the baryon number, namely, $B = 0$, the particle b has the baryon number $B = 1$. Thus, it can be understood that the variation of the baryon numbers is $\Delta B = 1$ in the process. What is important is that there is no baryon number violating process in the SM, and the process has not been seen in any experiments,

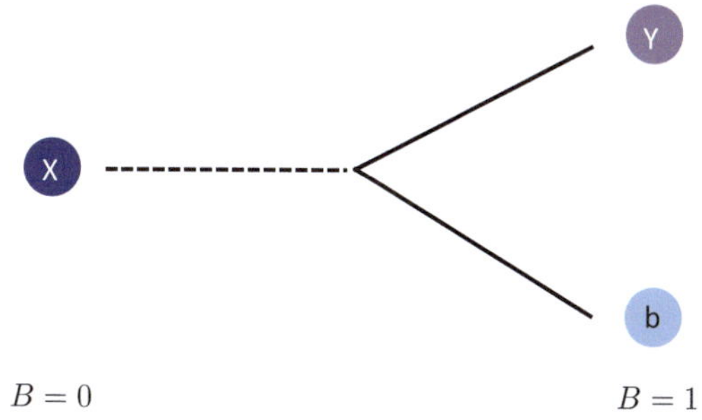

Fig. 1.2 An example of process which violates baryon numbers. Since particle X and Y do not possess baryon numbers, the variation of the baryon number amounts to $\Delta B = 1$

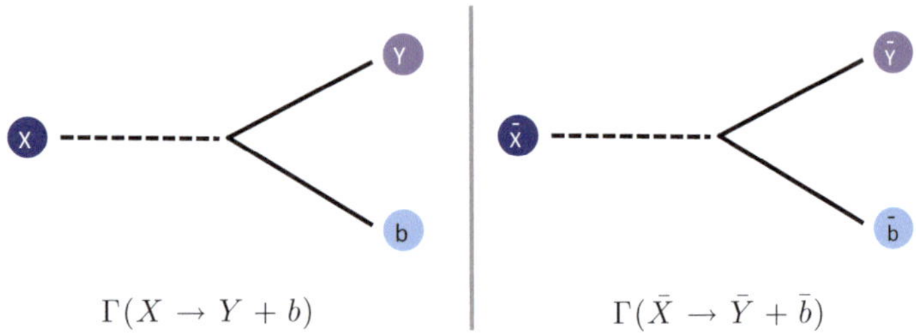

Fig. 1.3 Decay process of particle X and antiparticle $\bar{X}$. If $\Gamma(X \to Y+b) = \Gamma(\bar{X} \to \bar{Y}+\bar{b})$, the net baryon number equals zero

so far. In other words, the process is necessary for the BAU, but it must also evade the experimental constraint such as proton decay. As discussed later, one interesting process is that caused by the sphaleron [7, 8], which violates the $B + L$ number through anomaly. The sphaleron process frequently occurs at high temperature, but it is suppressed at zero temperature. So, we do not need to worry about the experimental limit on it.

The second condition naively implies that a difference of transition probability between $\Gamma(X \to Y + b)$ and $\Gamma(\bar{X} \to \bar{Y} + \bar{b})$ is needed, where the character with a bar is denoted as an antiparticle. In Fig. 1.3, while the left process creates $\Delta B = 1$, the right process results in $\Delta B = -1$. If $\Gamma(X \to Y+b) = \Gamma(\bar{X} \to \bar{Y}+\bar{b})$, the total BAU eventually vanishes. Thus, the difference of the transition probability between baryon and anti-baryon is also needed to leave the baryon asymmetry. Actually, the condition requires us to violate symmetry in which the baryon number is changed into odd number. Let us look into more details.

Once we define the baryon number B as follows

$$B = \sum_q i \int d^3x \, \frac{1}{3} \bar{q} \gamma^0 q, \tag{1.2}$$

with a quark field q, one can check that the sign changes under C and CP transformations. We also introduce density operator $\rho(t)$ to describe the state of the Universe

$$\rho(t) = \sum_n p_n |\psi_n(t)\rangle \langle \psi_n(t)|, \tag{1.3}$$

where $\{|\psi_n(t)\rangle\}$ is a complete set of states in the Schrödinger picture and p_n corresponds to a probability that a state of $|\psi(t)\rangle$ can be found. The evolution of the density operator is given by solving the Liouville equation

$$i\hbar \frac{\partial \rho(t)}{\partial t} + [\rho(t), H] = 0, \tag{1.4}$$

with the Hamiltonian H. Defining $L\rho(t) = [H, \rho(t)]$, we can obtain the formal solution for the density operator as $\rho(t) = \rho_0 e^{-\frac{i}{\hbar}Lt}$.[1] On the other hand, in thermal equilibrium, the solution is given by $\rho_{\rm eq} \equiv \exp(-\beta H)$ with $\beta = 1/T$.

With the density operator, an expectation value of an operator $\mathcal{O}$ can be derived as

$$\langle \mathcal{O} \rangle(t) = \mathrm{Tr}[\rho(t)\mathcal{O}]. \tag{1.5}$$

Here, let us assume that the Universe is initially baryon-symmetric, which implies that $\langle B \rangle(t = 0) \equiv \mathrm{Tr}[\rho_0 B] = 0$. If the Hamiltonian is invariant under C and CP transformations, namely,

$$[H, C] = [H, CP] = 0, \tag{1.6}$$

we obtain that

$$[\rho(t), C] = [\rho(t), CP] = 0, \tag{1.7}$$

and these relations lead to

$$\langle B \rangle(t) = \mathrm{Tr}[\rho(t)B] = -\mathrm{Tr}[\rho(t)CBC^{-1}] = -\mathrm{Tr}[\rho(t)B] = 0, \tag{1.8}$$

$$\langle B \rangle(t) = \mathrm{Tr}[\rho(t)B] = -\mathrm{Tr}[\rho(t)(CP)B(CP)^{-1}] = -\mathrm{Tr}[\rho(t)B] = 0. \tag{1.9}$$

Thus, it can be understood that the final baryon number disappear if both C and CP symmetries are kept. These discussions also imply that any symmetry under which the baryon number changes the sign must be broken to leave the nonzero value.[2]

Finally, the third condition represents that the transition probability of the baryon number violating process should be different from that of the reversal process. If $\Gamma(X \to Y + b) = \Gamma(Y + b \to X)$, the net baryon number finally becomes zero as illustrated in Fig. 1.4. Hence, the non-equilibrium is needed.

Each scenario introduced in the previous section does satisfy Sakharov's criteria, and among them is electroweak baryogenesis (EWBG) [9]. In the mechanism, the BAU can be produced during the electroweak phase transition (EWPT). Therefore the relevant energy scale would be within our experimental reach, i.e., $\mathcal{O}(100)$ GeV, which leads to the high testability by collider experiment. Actually, as discussed later, the possibility of the Standard Model EWBG has been completely excluded by the discovery of the Higgs boson. Moreover, not only the collider experiment but precision measurement of electric dipole moments (EDMs) can probe it. Actually,

[1] In the Heisenberg picture, it can be written that $\langle \mathcal{O} \rangle(t) = \mathrm{Tr}[\rho_0 \mathcal{O}(t)]$ with $\rho_0 = \rho(t = 0)$.

[2] Respecting the CPT theorem, one would obtain $[CPT, \rho_{\rm eq}] = 0$ that results from $[CPT, H] = 0$. Since time-reversal transformation does not change the sign of the baryon number, it leads to $(CPT)B(CPT)^{-1} = -B$. Thus, it is concluded that the baryon number cannot be produced. This situation represents that the expression of $\rho(t)$ in thermal equilibrium is not valid.

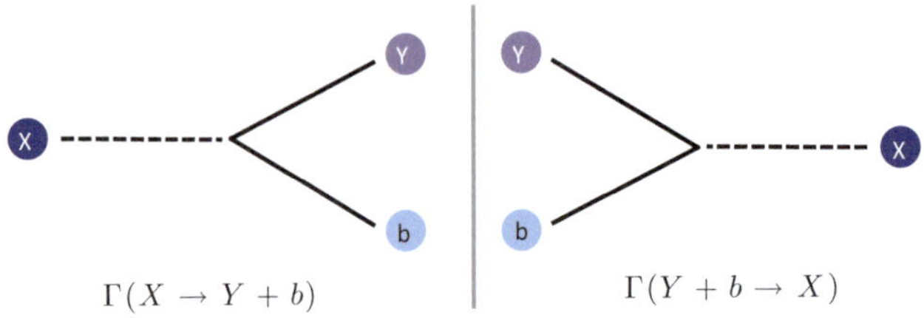

Fig. 1.4 The baryon number violating process of $X \to Y + b$ and the reversal process of $Y + b \to X$. If $\Gamma(X \to Y + b) = \Gamma(Y + b \to X)$, the baryon number vanishes even if it is produced once

it seems that the mechanism is more complicated than other scenarios, but the examination would be an urgent issue since the collider experiment is running now. For reviews on EWBG, see [3, 10–18].

1.3 Outline of This Thesis

In this thesis, we firstly review EWBG and give a comment on the current situation. The discussion about a model we focus on starts in Chap. 3, and we explain a calculation of the baryon number based on the closed time path formalism in Chap. 4. Subsequently, a relationship between the CP-violating source and EDMs in the model is discussed. Then, we show our numerical results and describe the verifiability. Finally, Chap. 7 is devoted to conclusions.

References

1. P.A.R. Ade et al. [Planck Collaboration], Astron. Astrophys. **571**, A16 (2014) [arXiv:1303.5076 [astro-ph.CO]]
2. G. Steigman, Ann. Rev. Astron. Astrophys. **14**, 339 (1976)
3. K. Funakubo, Prog. Theor. Phys. **96**, 475 (1996) [hep-ph/9608358]
4. Based on K. Funakubo's lecture slides held in the intensive course at Nagoya University on 20–21 June 2013
5. M. Shaposhnikov, J. Phys. Conf. Ser. **171**, 012005 (2009)
6. A.D. Sakharov, Pisma Zh. Eksp. Teor. Fiz. **5**, 32 (1967); [JETP Lett. **5**, 24 (1967)]; [Sov. Phys. Usp. **34**, 392 (1991)]; [Usp. Fiz. Nauk **161**, 61 (1991)]
7. N.S. Manton, Phys. Rev. D **28**, 2019 (1983). https://doi.org/10.1103/PhysRevD.28.2019
8. F.R. Klinkhamer, N.S. Manton, Phys. Rev. D **30**, 2212 (1984)
9. V.A. Kuzmin, V.A. Rubakov, M.E. Shaposhnikov, Phys. Lett. B **155**, 36 (1985)
10. A.G. Cohen, D.B. Kaplan, A.E. Nelson, Ann. Rev. Nucl. Part. Sci. **43**, 27 (1993) [hep-ph/9302210]
11. M. Quiros, hep-ph/9901312

12. V.A. Rubakov, M.E. Shaposhnikov, Usp. Fiz. Nauk **166**, 493 (1996); [Phys. Usp. **39**, 461 (1996)] [hep-ph/9603208]
13. M. Trodden, Rev. Mod. Phys. **71**, 1463 (1999) [hep-ph/9803479]
14. W. Bernreuther, Lect. Notes Phys. **591**, 237 (2002) [hep-ph/0205279]
15. J.M. Cline, hep-ph/0609145
16. D.E. Morrissey, M.J. Ramsey-Musolf, New J. Phys. **14**, 125003 (2012) [arXiv:1206.2942 [hep-ph]]
17. T. Konstandin, Phys. Usp. **56**, 747 (2013); [Usp. Fiz. Nauk **183**, 785 (2013)] [arXiv:1302.6713 [hep-ph]]
18. A. Riotto, hep-ph/9807454

Chapter 2
Electroweak Baryogenesis

Abstract The Higgs boson would be the key particle to EWPT, so the establishment of the Higgs sector plays an essential role in EWBG. In fact, the discovery of the Higgs boson has narrowed down the possibilities of EWBG in various models. The high testability is one reason why EWBG is attractive. Given that the Large Hadron Collider (LHC) is running now, we expect that the LHC can examine more feasible parameter region. This chapter describes how electroweak baryogenesis satisfies the Sakharov criteria and creates the baryon asymmetry. Subsequently, a current status of the scenario is mentioned.

Keywords Electroweak phase transition · Sphaleron process · New physics

2.1 Electroweak Baryogenesis

In EWBG, the Sakharov criteria are satisfied by the following processes:

1. Baryon number violating process: sphaleron process
2. C and CP violations: chiral gauge theory and CP phase
3. Out of equilibrium: first-order electroweak phase transition

The first condition is automatically satisfied by the sphaleron process without dependence on a model, but it relies on the model whether the second and third conditions are satisfied.

2.1.1 Sphaleron Process

2.1.1.1 The $(B + L)$-Violating Process

The sphaleron process is basically the $(B+L)$ violating process at the quantum level, and the quantum process is catalyzed by the transition between degenerate vacua. While the process frequently occurs at high temperature, it is drastically suppressed

© Springer Nature Singapore Pte Ltd. 2018

K. Fuyuto, *Electroweak Baryogenesis and Its Phenomenology*, Springer Theses,
https://doi.org/10.1007/978-981-13-1008-9_2

near zero temperature. Therefore, we do not need to worry about experimental limits such as proton decay. The subsequent discussion about the sphaleron process is based on Ref. [1].

If we consider the Lagrangian that is invariant under $SU(3)_C \times SU(2)_L \times U(1)_Y$, global $U(1)$ symmetries accidentally show up. The accidental symmetries lead to the conservations of baryon number B and lepton number L in the SM. However, in 1976, 't Hooft found that there is a non-perturbative effect that violates the $(B+L)$ number [2]. This effect is called instanton, and its transition rate is extremely small, $e^{-4\pi/\alpha_2} \sim 10^{-150}$, where $\alpha_2 = g_2^2/(4\pi)$ and g_2 is the $SU(2)_L$ gauge coupling. Subsequently, in 1984, Klinkhamer and Manton realized that, although instanton is the process at zero temperature, such an effect is enhanced at high temperature [3].

Let us see the following phase transformation with respect to the Dirac field $\psi(x)$

$$\psi(x) \to \psi'(x) = e^{i(a+b\gamma_5)\theta(x)}\psi(x). \tag{2.1}$$

Here, while $a = 1/3$ and $b = 0$ correspond to the baryon number transformation for quark field, $a = 1$ and $b = 0$ are equivalent to the lepton number transformation for lepton field. Under this transformation, two variations appear in the generating function of $Z(\eta, \bar{\eta}, J) = \int \mathcal{D}\psi\mathcal{D}\bar{\psi}\, \exp[S_0(\psi, \bar{\psi}, A^a)]$ where A^a is gauge field. One is in the action

$$S_0 \to S_0 + \delta S_0, \tag{2.2}$$

with

$$\delta S_0 = -\int d^4x \left[\bar{\psi}(x)m\left(e^{2ib\gamma_5\theta(x)} - 1\right)\psi(x) + \bar{\psi}(x)\gamma^\mu(a+b\gamma_5)\psi(x)\partial_\mu\theta(x)\right]. \tag{2.3}$$

The other is in the measure

$$\mathcal{D}\psi\mathcal{D}\bar{\psi} \to \mathcal{D}\psi\mathcal{D}\bar{\psi}\; S_1, \tag{2.4}$$

with

$$S_1 = \exp\left[i\int d^4x\,\theta(x)\left\{\frac{(a-b)}{8\pi^2}\mathrm{Tr}\left[F^{(L)\mu\nu}\tilde{F}^{(L)}_{\mu\nu}\right] - \frac{(a+b)}{8\pi^2}\mathrm{Tr}\left[F^{(R)\mu\nu}\tilde{F}^{(R)}_{\mu\nu}\right]\right\}\right], \tag{2.5}$$

where $\tilde{F}_{\mu\nu} \equiv \epsilon_{\mu\nu\rho\sigma}F^{\rho\sigma}/2$. Here, $F^{L,R}_{\mu\nu}$ are the field strengths of gauge field, which couple to the left-handed current, $j_L = \bar{\psi}_L\gamma_\mu\psi_L$, and the right-handed current,

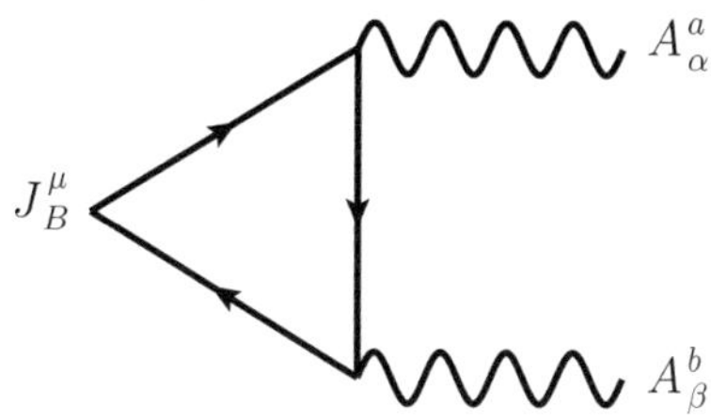

Fig. 2.1 Quantum effect that violates the $(B+L)$ number

$j_R = \bar{\psi}_R \gamma_\mu \psi_R$, respectively.[1] A requirement that the generating function does not change under the transformation yields

$$\partial^\mu \left\{\bar{\psi}(x)\gamma_\mu(a+b\gamma_5)\psi(x)\right\}$$
$$= -2ibm\bar{\psi}(x)\gamma_5\psi(x) - i\frac{(a-b)}{8\pi^2}\mathrm{Tr}\left[F^{(L)\mu\nu}\tilde{F}^{(L)}_{\mu\nu}\right] + i\frac{(a+b)}{8\pi^2}\mathrm{Tr}\left[F^{(R)\mu\nu}\tilde{F}^{(R)}_{\mu\nu}\right]. \tag{2.6}$$

The first term on the right-hand side is originated from the variation at the classical level, and it becomes zero as long as $b = 0$. However, the second and third terms do not vanish even if $b = 0$, which results from the quantum effect in Fig. 2.1. The situation indicates that, although the baryon and lepton number violation are prohibited at the classical level, they occur at the quantum level. Performing trace calculation, derivative of the baryon number current J^μ_B and the lepton number current J^μ_L are given by

$$\partial_\mu J^\mu_B = \sum_q \partial_\mu\left(\frac{1}{3}\bar{q}\gamma^\mu q\right) = i\frac{N_f}{32\pi^2}\left(-g_2^2 F^{a\mu\nu}\tilde{F}_{a\mu\nu} + g_1^2 B^{\mu\nu}\tilde{B}_{\mu\nu}\right), \tag{2.7}$$

$$\partial_\mu J^\mu_L = \sum_l \partial_\mu\left(\bar{l}\gamma^\mu l + \bar{\nu}_l\gamma^\mu\nu_l\right) = i\frac{N_f}{32\pi^2}\left(-g_2^2 F^{a\mu\nu}\tilde{F}_{a\mu\nu} + g_1^2 B^{\mu\nu}\tilde{B}_{\mu\nu}\right), \tag{2.8}$$

where N_f is the number of flavor and g_1 is $U(1)_Y$ gauge coupling. $F^a_{\mu\nu}$ $(a = 1, 2, 3)$ and $B^{\mu\nu}$ are the field strengths of $SU(2)$ gauge field A^a_μ and $U(1)$ gauge field B_μ, respectively. What is important is that the two relations imply $\partial_\mu(J^\mu_B - J^\mu_L) = 0$, but $\partial_\mu(J^\mu_B + J^\mu_L) \neq 0$. This is called $U(1)_{B+L}$ anomaly.

Again, once we introduce the baryon number denoted as

$$B = i\int d^3x\, J^0_B = \sum_q i\int d^3x\, \frac{1}{3}\bar{q}\gamma^0 q, \tag{2.9}$$

[1]Chiral fermions are defined as $\psi_L(x) = P_L\psi(x)$ and $\psi_R(x) = P_R\psi(x)$ with projection operators, $P_L = (1-\gamma_5)/2$ and $P_R = (1+\gamma_5)/2$.

the time variation ΔB can be expressed by those of the Chern-Simons numbers N_{CS} for $SU(2)_L$ and n_{CS} for $U(1)_Y$

$$\Delta B = B(t_f) - B(t_i) = N_f \left(\Delta N_{\text{CS}} - \Delta n_{\text{CS}}\right), \tag{2.10}$$

with

$$N_{\text{CS}} = -\frac{g_2^2}{16\pi^2}\int d^3x\, 2\epsilon^{ijk}\text{Tr}\left[\partial_i A_j A_k + i\frac{2}{3}g_2 A_i A_j A_k\right], \tag{2.11}$$

$$n_{\text{CS}} = -\frac{g_1^2}{16\pi^2}\int d^3x\, \epsilon^{ijk}\partial_i B_j B_k, \tag{2.12}$$

where $A_i \equiv A_i^a \tau^a/2$ and τ^a is the Pauli matrix.

Under the gauge transformations with representations of U_Y for $U(1)$ and U for $SU(2)$, the gauge fields A_i and B_i transform

$$B_i \to B_i + \frac{i}{g_1}(\partial_i U_Y)U_Y^{-1}, \tag{2.13}$$

$$A_i \to U A_i U^{-1} + \frac{i}{g_2}(\partial_i U)U^{-1}. \tag{2.14}$$

Thus, it is seen that, although n_{CS} does not change under the gauge transformation, N_{CS} does:

$$\delta N_{\text{CS}}(U) = \frac{1}{24\pi^2}\int d^3x\, \epsilon^{ijk}\text{Tr}\left[(\partial_i U)U^{-1}(\partial_j U)U^{-1}(\partial_k U)U^{-1}\right]. \tag{2.15}$$

Recalling Eq. (2.10), it can finally be understood that the variation of the baryon and lepton numbers are proportional to that of N_{CS}.

Generally, a 2×2 unitary matrix can be written as $U = a\mathbf{1} + i\mathbf{b}\cdot\boldsymbol{\tau}$ with $a^2 + |\mathbf{b}|^2 = 1$. Therefore, the topology of $SU(2)$ is equal to the three-dimensional sphere S_3. The property of S_3 is mathematically $\pi_3(S_3) = Z$,[2] which implies that multiple degenerate vacua distinguished by an integer exist. This situation results in the integer value of N_{CS}. For instance, if we take $U^{(1)}(x) = (x_0 + i\mathbf{x}\cdot\tau)/r$ with $r = (x_0^2 + |\mathbf{x}|^2)^{1/2}$, it leads to $\delta N_{\text{CS}}\left(U^{(1)}\right) = 1$. Moreover, taking $U^{(n)}(x) = \left[U^{(1)}\right]^n$ produces $\delta N_{\text{CS}}\left(U^{(n)}\right) = n$. Therefore, only the variation of the Chern-Simons number of $SU(2)$ can contribute to those of baryon and lepton numbers. However, the contribution to ΔB from $U(1)$ is nothing because $U(1)$ does not own degenerate vacua that topologically vary.[3] Thus, only left-handed particles are produced by the $(B + L)$-violating process.

[2] π_m implies the m-th homotopy group.

[3] Because, $\pi_3(U(1)) = 0$.

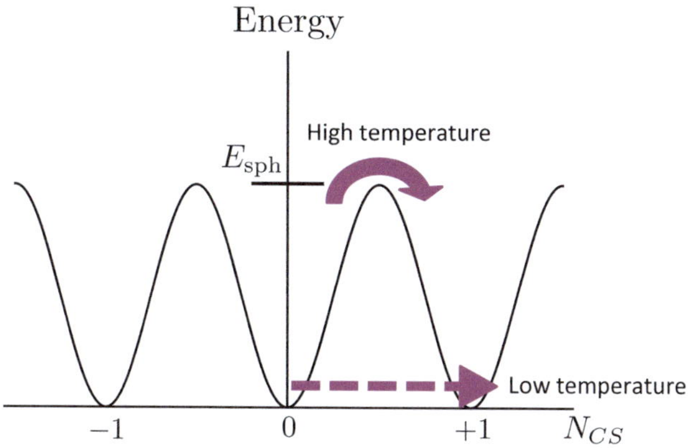

Fig. 2.2 Description of transition between two degenerate vacua implying. The curved arrow indicates the sphaleron process. The dashed arrow shows the instanton. The height of the potential corresponds to the energy of the sphaleron $E_{\rm sph}$

As mentioned in the opening sentence, the sphaleron process can be described by a transition between degenerate vacua characterized by $N_{\rm CS}$. Figure 2.2 shows two kinds of the transition that changes $N_{\rm CS}$. The lower dashed arrow indicates the instanton,[4] which occurs at low temperature. The transition rate is extremely suppressed and negligible [2]. On the other hand, the upper curved arrow represents the sphaleron, and it crosses the top of a barrier whose value corresponds to the energy of the sphaleron ($\sim$10TeV). Apart from the instanton process, the sphaleron process frequently occurs at high temperature. However, as discussed below, the rate differs between EW symmetric and broken phase.

Finally, we give a brief explanation about particle production through the sphaleron process. An intuitive explanation is held in Ref. [4]: vacuum is filled with particles that have negative energies. However, once particle interacts with gauge field, it can affect the energy shift of the particle. And, it follows that the production of the particle with positive energy. For the detailed discussion, see Ref. [5–7].

2.1.1.2 Sphaleron

Originally, the sphaleron implies a solution to the classical field equation at finite temperature. While the instanton is the classical solution in the 4-dimensional Euclidean action, the sphaleron is the static but unstable solution found by

[4]In a precise sense, this is called a constrained instanton.

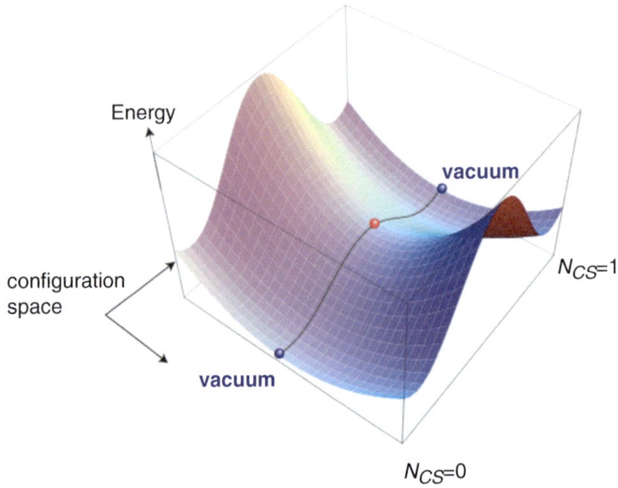

Fig. 2.3 Configuration of the sphaleron which is located at the saddle point in the least-energy path between two degenerate vacua [8]. (Copyright ©Koichi Funakubo)

Klinkhamer and Manton [3, 9]. As drawn in Fig. 2.3, the sphaleron corresponds to the configuration with $N_{\rm CS} = 1/2$ in the least-energy path that connect two distinct vacua next to each other. The configuration is located at the top of the barrier but a saddle point, and its energy is finite.

As carefully explained in [7, 8], the sphaleron solution can be found by a so-called minimax procedure. For finding the solution, it would be important whether a configuration that yields finite energy can be defined at the 3-dimensional infinity S_2. In the procedure, considering a product $S_1 \times S_2$, we firstly construct a 3-dimensional sphere S_3 described by (μ, θ, ϕ), where μ corresponds to a loop parameter with $\mu \in [0, \pi]$. The topology is $\pi_3(S^3) \simeq Z$, which implies an existence of noncontractible configuration. Along the direction of μ, we find a saddle point at $\mu = \pi/2$ with the most energy in the least path as in Fig. 2.3, and it corresponds to the sphaleron configuration. On another front, $\mu = 0$ and π corresponds to the vacuum configuration. Therefore, finding an extremum of maximum value of the sphaleron energy finally leads to the solution.

Let us take a look at 4-dimensional $SU(2)$ gauge-Higgs system. The Lagrangian is given by

$$\mathcal{L} = -\frac{1}{4} F^a_{\mu\nu} F^{a\mu\nu} + (D_\mu \Phi)^\dagger D^\mu \Phi - \lambda \left(\Phi^\dagger \Phi - \frac{v^2}{2} \right)^2, \tag{2.16}$$

where Φ is the $SU(2)$ doublet Higgs field and $v \simeq 246\,\text{GeV}$. Taking $A_0 = 0$ gauge, one can obtain a static energy[5]

$$E_{\text{sph}} = \int d^3x \left[\frac{1}{4} F^a_{ij} F^a_{ij} + (D_i\Phi)^\dagger (D_i\Phi) + \lambda \left(\Phi^\dagger \Phi - \frac{v^2}{2} \right)^2 \right]. \tag{2.17}$$

In order to obtain the finite E_{sph}, the fields Φ and A_i should be the vacuum configurations at $r = |x| = \infty$

$$\Phi(r = \infty) = U(\theta, \phi) \begin{pmatrix} 0 \\ v/\sqrt{2} \end{pmatrix}, \qquad A_i(r = \infty) = -\frac{i}{g_2} \partial_i U(\theta, \phi) U^{-1}(\theta, \phi), \tag{2.18}$$

where spherical coordinate (r, θ, ϕ) is employed. Let us extend $U(\theta, \phi)$ to a noncontractible loop configuration $U(\mu, \theta, \phi)$

$$U(\mu, \theta, \phi) = \begin{pmatrix} e^{i\mu}(\cos\mu - i \sin\mu \cos\theta) & e^{i\phi} \sin\mu \sin\theta \\ -e^{-i\phi} \sin\mu \sin\theta & e^{-i\mu}(\cos\mu + i \sin\mu \cos\theta) \end{pmatrix}, \tag{2.19}$$

where it is required that $U(\mu, \theta, \phi) = U(\mu, \theta + \pi, \phi) = U(\mu, \theta, \phi + 2\pi)$ and $U(0, \theta, \phi) = U(\pi, \theta, \phi) = 1$. With the above configuration, the ansatz can be described by [3, 9, 10]

$$\Phi(\mu, r, \theta, \phi) = \frac{v}{\sqrt{2}} \left[(1 - h(r)) \begin{pmatrix} 0 \\ e^{-i\mu} \cos\mu \end{pmatrix} + h(r) U(\mu, \theta, \phi) \begin{pmatrix} 0 \\ 1 \end{pmatrix} \right], \tag{2.20}$$

$$A_i(\mu, r, \theta, \phi) = -\frac{i}{g_2} f(r) \partial_i U(\mu, \theta, \phi) U^{-1}(\mu, \theta, \phi). \tag{2.21}$$

The functions of f and h imposed the following boundary conditions on

$$\lim_{\xi \to 0} f(\xi) = 0, \qquad \lim_{\xi \to 0} h(\xi) = 0, \tag{2.22}$$

$$\lim_{\xi \to \infty} f(\xi) = 1, \qquad \lim_{\xi \to \infty} h(\xi) = 1. \tag{2.23}$$

[5]Here, we neglect a contribution from $U(1)_Y$ since the effect is a few %.

with the dimensionless radial distance $\xi \equiv g_2 v r$. These ansatz yield

$$E_{\rm sph} = \frac{4\pi v}{g_2}\int_0^\infty d\xi \, \sin^2\mu \left[\left\{4\left(\frac{df}{d\xi}\right)^2 + \frac{2}{\xi^2}[f(1-f)]^2\sin^2\right\} + \frac{\xi^2}{2}\left(\frac{dh}{d\xi}\right)^2 \right.$$
$$+[h(1-f)]^2 + (1-h)^2 f^2\cos^2\mu - 2h(1-h)f(1-f)\cos^2\mu$$
$$\left. + \frac{\lambda}{4g_2^2}\xi^2(h^2-1)^2\sin^2\mu\right]. \tag{2.24}$$

It should be emphasized that the sphaleron energy functional is proportional to v. In other words, the sphaleron energy becomes 0 if in the symmetric phase, that is, the sphaleron solution exists only in the broken phase. Again, as mentioned before, the $(B+L)$-violating process can occur in the symmetric phase, and we generically call it the sphaleron process.

Taking $\mu = \pi/2$, one can obtain the equations of motion for the sphaleron solution

$$\frac{d^2}{d\xi^2}f(\xi) = \frac{2}{\xi^2}f(\xi)(1-f(\xi))(1-2f(\xi)) - \frac{1}{4}h^2(\xi)(1-f(\xi)), \tag{2.25}$$

$$\frac{d}{d\xi}\left(\xi^2\frac{dh(\xi)}{d\xi}\right) = 2h(\xi)(1-f(\xi))^2 + \frac{\lambda}{g_2^2}\xi^2(h^2(\xi)-1)h(\xi). \tag{2.26}$$

In the equations, only λ and g_2 can enter as parameters. Since we know not only g_2 but λ by the measurements of the Higgs mass, the sphaleron energy is finally given by $\sim$ 9.08 TeV [11]. It should be emphasized that the sphaleron energy estimated with finite temperature potential becomes smaller than that with only zero-temperature potential, because the true solution is obtained by including the finite-temperature effect. Moreover, it differs in each model since it depends on the Higgs potential as estimated in [12–14]. The value is crucially important for a criterion of the strong first-order phase transition. Table 2.1 shows some values of $\mathcal{E}(T)$ normalized by $E_{\rm sph}(T) = 4\pi v(T)\mathcal{E}(T)/g_2$.

One may wonder whether the sphaleron process can be observed in high-energy collisions. Actually, it has been a long-standing question [15–18]. Recently, the discussion has been revisited by [19], in which they claim that band effect caused by

Table 2.1 Sphaleron energy in various models. The sphaleron energy is normalized by $E_{\rm sph}(T) = 4\pi v(T)\mathcal{E}(T)/g_2$, and T_C represents critical temperature at which two degenerate minima appear in the Higgs potential

Model	$\mathcal{E}(T_C)$
Standard model [3]	2.0
Minimal supersymmetric standard model [12]	1.769
Real-singlet model [13]	1.92

periodic potential of the sphaleron can enhance the process yielding detectable cross section for the $(B + L)$-violating events. However, as discussed in [20, 21], when we formulate the $(B + L)$-changing scattering amplitude, there would be another suppression factor coming from an overlap between the state for the classical configuration and the physical two-particle states.

2.1.2 *Chiral Gauge Theory and CP Phase*

If theory possesses chiral interaction and CP phase, it implies C and CP violations. Therefore, the theory can satisfy the Sakharov second condition. Actually, the SM can satisfy it, because the SM is the chiral gauge theory and it maximally breaks C symmetry. Besides, the CP phase known as the Kobayashi-Maskawa (KM) phase also exists in the model.

2.1.3 *First-Order Electroweak Phase Transition*

In electroweak phase transition (EWPT), the vacuum expectation value (VEV) of the $SU(2)$ Higgs doublet is an order parameter, and the first-order EWPT implies that the VEV suddenly becomes nonzero value at some critical temperature T_C as seen in Fig. 2.4. Although the VEV is zero when the Universe is at high temperature, it suddenly gets a nonzero value v_C at T_C. This discontinuity is the remarkable feature of the first-order EWPT, while the critical VEV does not exist in the second-order EWPT.

In Fig. 2.5, the situation at the first-order EWPT is described in terms of the effective potential of the Higgs. At $T > T_C$, the electroweak symmetry is kept, and then, two degenerate minima appear at T_C. As the Universe cools, the effective potential eventually takes the wine-bottle shape. What is important is that a negative contribution is necessary to induce the degenerate minima in the effective potential, and as discussed later, this contribution comes from only bosons.

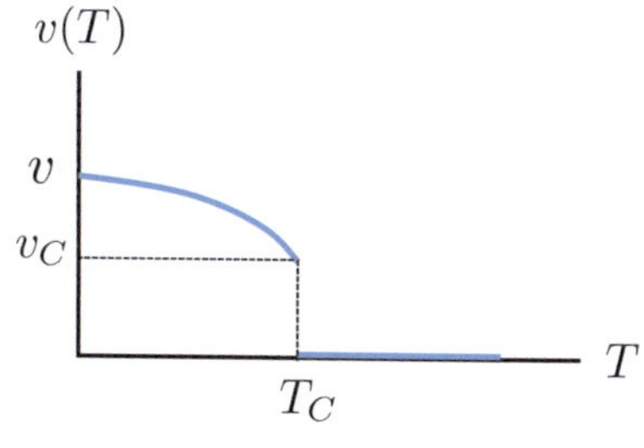

Fig. 2.4 Dependence on the VEV of the Higgs doublet on temperature at the first-order EWPT. Although $v = 0$ at high temperature, it gets nonzero value denoted as v_C at the critical temperature T_C. And then, $v(T)$ eventually becomes that at zero temperature as temperature decreases

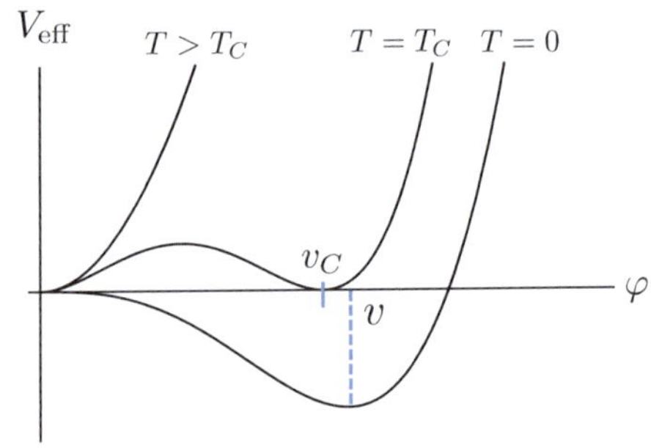

Fig. 2.5 Shape of the effective potential $V_{\rm eff}$ of the Higgs in the first-order EWPT. At $T > T_C$, it has only one minimum, and then, two degenerate vacua separated by a potential barrier appear at $T = T_C$. Finally, the shape becomes that at zero temperature

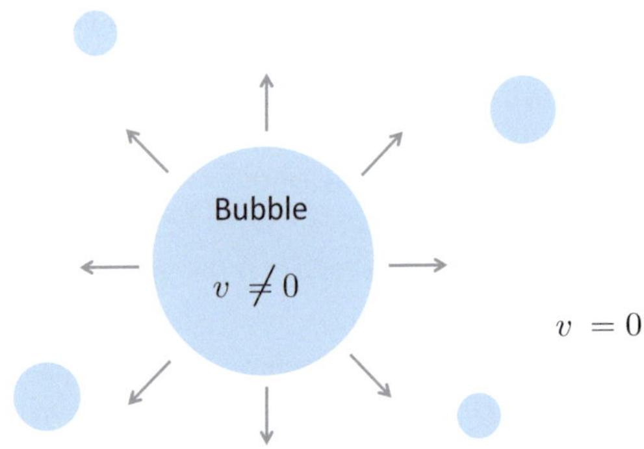

Fig. 2.6 Description of a bubble nucleation. Bubble wall separates the broken phase ($v \neq 0$) from the symmetric phase ($v = 0$). Once the bubbles are created, they start to expand. EWPT finishes when the Universe is filled with these bubbles

If the first-order EWPT is achieved, a bubble can be nucleated as shown Fig. 2.6. One may easily be able to imagine the situation by considering boiled water in which two different phases, that is, liquid and gas, exist. The EW symmetry is broken inside the bubble, but the symmetry is kept outside it. Once the bubbles are nucleated, they start to expand. As mentioned in the beginning of the section, the third condition of the Sakharov criteria can be satisfied by the expansion of the bubble, because the $(B + L)$-changing rates in the broken and symmetric phases are different from each other. That is, the bubble expansion makes it possible to store particles, which are produced by the sphaleron process in the symmetric phase, before the inverse process occurs.

In Fig. 2.7, a point described by a cross mark is in the symmetric phase at $t = t_1$. As a bubble expands, the cross mark is included in the bubble at $t = t_2$ $(> t_1)$. The rates in the broken $(t = t_2)$ and symmetric $(t = t_1)$ phases are given by

$$\Gamma_{\rm sph}^{(b)} \simeq T^4 e^{-E_{\rm sph}/T}, \tag{2.27}$$

$$\Gamma_{\rm sph}^{(s)} \sim \kappa(\alpha_2 T)^4, \tag{2.28}$$

where (b) and (s) represent the broken and the symmetric phase, respectively. While the former is calculated by the WKB approximation [22], and the latter is estimated

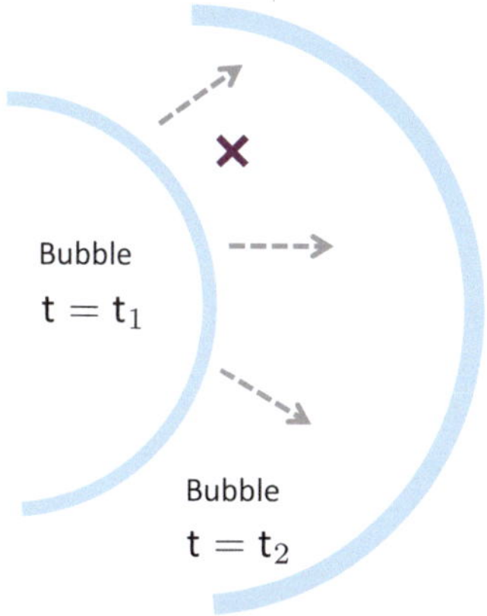

Fig. 2.7 Description of a bubble expansion. At $t = t_1$, a point expressed by a cross mark is in the symmetric phase. As the bubble expands, it can be included inside the bubble in which the electroweak symmetry is breaking

by dimensional analysis with $\kappa = 0.1 \sim 1.0$ [23, 24].[6] As seen in these expressions, the $(B + L)$-changing rate in the broken phase is extremely suppressed by the Boltzmann suppression factor, but that in the symmetric phase does not receive such a suppression.

In order to include the produced particles by the sphaleron process inside bubble, its expansion speed would be important. If it is faster than the sphaleron process in the symmetric phase, nonzero baryon number can be successfully incorporated in the bubble before the inverse reaction occurs.

Based on discussions in [8, 25], let us estimate time scales of relevant interaction to relativistic particle at temperature T. The time scale is naively given by the mean free path λ

$$t \simeq \lambda = \frac{1}{\sigma \cdot n(T)}, \tag{2.29}$$

where $\sigma \simeq \alpha^2/T^2$ is total cross section and $n(T) \simeq g_* \zeta(3) T^3/\pi^2$ is the particle number density with the Riemann zeta function $\zeta(3) = 1.2020569\cdots$, and the effective degree of freedom $g_* = \sum_{\text{boson}} g_B + \frac{3}{4}\sum_{\text{fermi}} g_F$. Taking $T \sim T_C \sim 100$ GeV, one can obtain the following time scales

$$t_H = H(T)^{-1} \simeq \frac{m_{\text{pl}}}{1.66\sqrt{g_*}T^2} \simeq 10^{14}\ \text{GeV}^{-1}, \qquad \text{(Hubble)}$$

$$t_s \simeq \frac{1}{\alpha_s^4 T} \simeq 0.1\ \text{GeV}^{-1}, \qquad \text{(QCD interaction)}$$

[6]The Monte Carlo simulation is used in Ref. [23], and lattice simulations with hard-thermal loop are performed in Ref. [24].

$$t_W \simeq \frac{1}{\alpha_2^4 T} \simeq 1\ \text{GeV}^{-1}, \qquad \text{(Weak interaction)}$$

$$t_{\text{sph}}^{(s)} \simeq \frac{1}{\alpha_2^4 T} \simeq 10^3\ \text{GeV}^{-1}, \qquad \text{(Sphaleron process)}$$

where we use the Hubble parameter as $H(T) \simeq 1.66\sqrt{g*}T^2/m_{\text{pl}}$ with the Planck mass $m_{\text{pl}} = 1.2 \times 10^{19}$ GeV. The above expressions say that $t_s < t_W < t_{\text{sph}}^{(s)} \ll t_H$, which implies that all of the interactions are chemical equilibrium. On the other hand, the wall width L_w and the velocity v_w of the bubble are estimated in [26, 27]

$$L_w \simeq 0.01 \sim 0.4\ \text{GeV}^{-1}, \qquad v_w \simeq 0.1c \sim 0.8c, \tag{2.30}$$

where c is the speed of light. Thus, the time scale of the bubble expansion is given by

$$t_w = \frac{L_w}{v_w} \simeq (0.001 \sim 4)\ \text{GeV}^{-1} \ll t_{\text{sph}}^{(s)}. \tag{2.31}$$

It turned out that the sphaleron process is in nonequilibrium, and the dynamics of the bubble makes it possible to capture the produced particles before the sphaleron inverse process.

2.2 Mechanism of Electroweak Baryogenesis

As we have seen so far, if electroweak phase transition is the first order, bubble can be nucleated at nucleation temperature T_N. T_N is generally lower than T_C,[7] and this situation is the so-called supercooling. The bubbles expand and the Universe is finally filled with them, which implies EWPT ends.

Once the bubble is nucleated, particle and antiparticle near the bubble are scattered as in Fig. 2.8, where the particles and the antiparticles are described with the chirality indexes L and R. In the scattering process, CP violation yields a difference of transmission to the bubble between the particle and antiparticle. Figure 2.9 shows an example where the antiparticles are bounced while the particles are transmitted. And, it follows that number density between them also differ. Defining the number density of the particle and antiparticle as $n_{L,R}$ and $n_{\bar{L},\bar{R}}$, one can understand the difference near the bubble as

$$n_L - n_{\bar{L}} \neq 0, \qquad n_R - n_{\bar{R}} \neq 0. \tag{2.32}$$

[7]For example, $T_N = 77.8$ GeV and $T_C = 91.5$ GeV in the scale-invariant two Higgs doublet model [14].

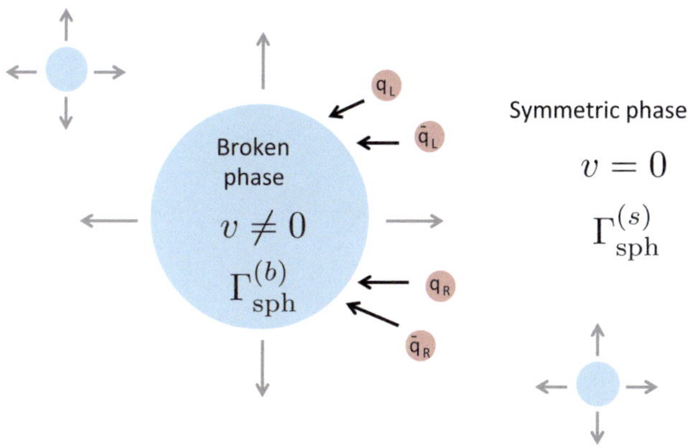

Fig. 2.8 As the bubble expands, particle q and antiparticle $\bar{q}$ collide with it

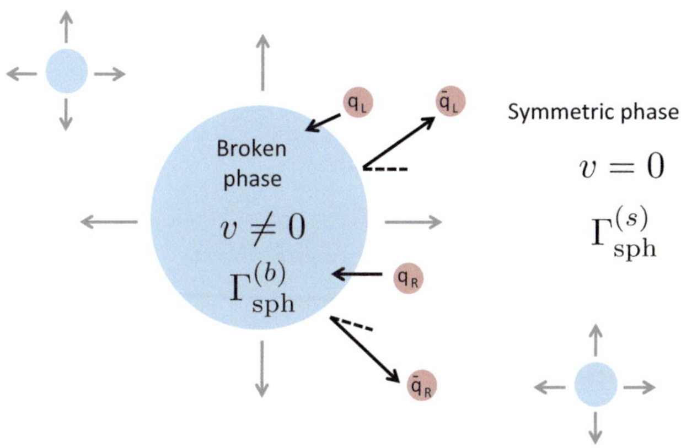

Fig. 2.9 If theory possesses CP violation, transmission rate to the bubble is different between the particle and antiparticle. This figure shows an example that the antiparticles $\bar{q}_L$ and $\bar{q}_R$ are bounced

At this point, however, the net baryon number density still remains zero

$$n_B = n_L - n_{\bar{L}} + n_R - n_{\bar{R}} = 0. \tag{2.33}$$

Once the sphaleron process occurs, the above expression can be changed because the left-handed particles are produced by it, which results in[8]

$$n_B = n_L - n_{\bar{L}} + n_R - n_{\bar{R}} \neq 0. \tag{2.34}$$

[8]Here, although we simply consider only the sphaleron process, there are actually other processes involved in changing the number densities such as the Yukawa interaction. If such an interaction is in equilibrium, it works to decrease the produced left-handed particles.

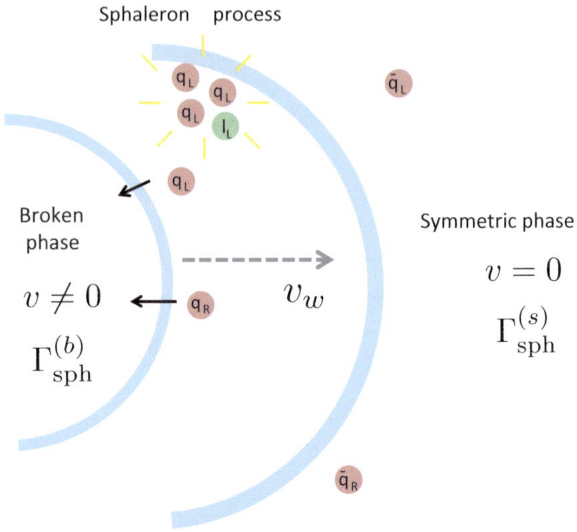

Fig. 2.10 The produced $(B + L)$ numbers can be included in the bubble thanks to its expansion right after the sphaleron process

Then, the nonzero n_B can finally be included into the bubble as in Fig. 2.10.[9]

2.2.1 *Strong First-Order Electroweak Phase Transition*

Actually, even if the bubble takes the nonzero baryon number in, it may disappear if the sphaleron process inside the bubble is enough fast to wash out it. In order to avoid this situation, we need to impose one important condition on the sphaleron rate in the broken phase

$$\Gamma_{\rm sph}^{(b)} < H(T), \tag{2.35}$$

and this expression can be rewritten by [12]

$$\begin{aligned} \frac{v(T)}{T} > \frac{g_2}{4\pi\mathcal{E}(T)} \Bigg[& 42.97 + \log(\kappa\mathcal{N}_{\rm tr}\mathcal{N}_{\rm rot}) + \log\left(\frac{\omega_-}{m_W}\right) - \frac{1}{2}\log\left(\frac{g_*}{106.75}\right) \\ & -2\log\left(\frac{T}{100\ \mathrm{GeV}}\right)\Bigg] \\ \equiv \zeta_{\rm sph}(T), & \end{aligned} \tag{2.36}$$

[9]For simplicity, only three left-handed quarks and left-handed lepton are described in Fig. 2.10, but the quarks and leptons over three generations are actually produced in the sphaleron process.

Table 2.2 The baryon number conservation condition in three models: minimal supersymmetric standard model [12], real-singlet model [13], and scale-invariant two Higgs doublet model [14]

Model	v_C/T_C	$\zeta_{\rm sph}(T_C)$	v_N/T_N	$\zeta_{\rm sph}(T_N)$
Minimal supersymmetric standard model	$\frac{107.096}{116.274} = 0.921$	–	$\frac{116.727}{115.585} = 1.010$	1.383
Real-singlet model	$\frac{172.83}{148.87} = 1.16$	1.14	–	–
Scale-invariant two Higgs doublet model	$\frac{211.0}{77.8} = 2.31$	1.23	$\frac{229.0}{77.8} = 2.94$	1.20

where $\mathcal{N}_{\rm tr}$ and $\mathcal{N}_{\rm rot}$ are the translational and rotational zero-mode factors of the fluctuations about the sphaleron, ω_- is the negative mode, and κ is the $\mathcal{O}(1)$ factor [28, 29]. This criterion is called the baryon number conservation condition, and the situation that the condition is satisfied is known as the strong first-order EWPT. The dominant contribution to the right side comes from $\mathcal{E}(T)$, and the large value results in weak condition for the strong first-order EWPT.

The baryon number conservation criterion is usually estimated at T_C or T_N, namely, it would be $v_C/T_C > \zeta_{\rm sph}(T_C)$ or $v_N/T_N > \zeta_{\rm sph}(T_C)$. v_N is the VEV at T_N which is larger than v_C; therefore it is obtained that $v_N/T_N > v_C/T_C$. It should be emphasized that, although many literatures roughly use $v_C/T_C > 1.0$, the right and left sides of the condition should independently be estimated and compared. The condition originally differs in each model and it does affect a possible range of model parameter. As discussed later, the baryon number conservation condition is really important in terms of testability of EWBG, because it can predict physical quantities such as mass of a new particle and coupling. Table 2.2 shows v_C/T_C, $\zeta_{\rm sph}(T_C)$, v_N/T_N, and $\zeta_{\rm sph}(T_N)$ in various models [12–14], and the unit of v and T is GeV. These analyses have been done with potential composed of tree, the one-loop Coleman-Weinberg [30, 31] and the finite-temperature one-loop potential [32]. Besides, the contribution of the zero mode factors are included in [12], and it is referred that the effect amounts to 10%. For the derivation of $\mathcal{N}_{\rm tr}\mathcal{N}_{\rm rot}$, see [12].

2.2.2 *Bubble Nucleation*

Once the bubble is nucleated at T_N, EWPT can start. The bubble can grow if the radius is larger than some critical size and convert fill the Universe to the broken phase, finally. The bubble with the critical size is called the critical bubble, and it can be formulated when a certain region simultaneously has a nonzero VEV.

As discussed in Ref. [33], the critical radius of the bubble is extracted by the nucleation rate per unit time and unit volume

$$\Gamma_N(T) = \Gamma_0 e^{-\Delta F/T}, \tag{2.37}$$

where Γ_0 is a prefactor and ΔF represents the variation of the free energy with a radius r

$$\Delta F = \frac{4\pi}{3}\left[p_s(T) - p_b(T)\right] r^3 + 4\pi r^2 \sigma, \tag{2.38}$$

where $p_{s(b)}$ corresponds to the pressure in the symmetric (broken) phase and σ is the surface free energy. They are given by $p_s(T) = -V_{\rm eff}(0; T)$, $p_b(T) = -V_{\rm eff}(v(T); T)$ and $\sigma = \int dz\ (\partial\varphi/\partial z)^2$ with the constant background field φ of the $SU(2)$ Higgs and perpendicular coordinate to the bubble wall. It depends on the size of the radius whether the bubble shrinks or expands: if the radius is small, the surface free energy dominates, but the pressure does if the radius is large. The radius that balances these two contributions is called the critical radius r_*, which is given by considering a maximum of ΔF

$$r_* = \frac{2\sigma}{p_b(T) - p_s(T)}. \tag{2.39}$$

The bubble can grow if $r > r_*$, and the expression yields $\Delta F_* = 4\pi\sigma r_*^2/3$.

The nucleation temperature is defined by [34]

$$\Gamma_N(T_N)/H^3(T_N) = H(T_N), \tag{2.40}$$

where $\Gamma_N(T_N)$ is the nucleation rate

$$\Gamma_N(T) \simeq T^4 \left(\frac{E_{\rm cb}(T)}{2\pi T}\right)^{\frac{3}{2}} e^{-\frac{E_{\rm cb}(T)}{T}}, \tag{2.41}$$

with the energy of the critical bubble $E_{\rm cb}(T)$. The Eq. (2.40) implies that one bubble can be nucleated in the horizon at T_N, and it follows that

$$\frac{E_{\rm cb}(T_N)}{T_N} - \frac{3}{2}\log\left(\frac{E_{\rm cb}(T_N)}{T_N}\right) = 152.59 - 2\log g_*(T_N) - 4\log\left(\frac{T_N}{100\ {\rm GeV}}\right). \tag{2.42}$$

Thus, it is seen that $E_{\rm cb}/T \lesssim 150$ is naively necessary for the EWPT.

The critical bubble is the static solutions for equations of motion with the least energy, which smoothly connects the symmetric phase to the broken phase. The solutions have been obtained in a concrete model such as the MSSM [12] and scale-invariant 2HDM [14], numerically. For example, the energy of the critical bubble in the scale-invariant 2HDM is described by

$$E_{\rm cb} = \int d^3x \left[(\partial_i\Phi_1)^\dagger\partial_i\Phi_1 + (\partial_i\Phi_2)^\dagger\partial_i\Phi_2 + V_{\rm eff}(\Phi_1, \Phi_2, T)\right] \tag{2.43}$$

where Φ_i $(i = 1,\ 2)$ is the $SU(2)$ Higgs doublet and the pure-gauge configurations are employed for the gauge fields, yielding $F^a_{ij} = B_{ij} = 0$. Let us parametrize the Higgs fields as

$$\Phi_1(r) = \frac{1}{\sqrt{2}}\begin{pmatrix} 0 \\ \rho_1(r)\end{pmatrix}, \qquad \Phi_2(r) = \frac{1}{\sqrt{2}}\begin{pmatrix} 0 \\ \rho_2(r)\end{pmatrix} \tag{2.44}$$

with $r = |x|$, and then, we take the following dimension less quantities

$$\xi = v(T)r, \qquad h_1(\xi) = \frac{\rho_1(r)}{v(T)\cos\beta}, \qquad h_2(\xi) = \frac{\rho_2(r)}{v(T)\sin\beta}, \tag{2.45}$$

with $v_2/v_1 = \tan\beta$. Thus, the equations of motion for the bubble solution are given by

$$-\frac{1}{\xi^2}\frac{d}{d\xi}\left(\xi^2\frac{dh_{1(2)}}{d\xi}\right) + \frac{1}{v^4(T)\cos^2\beta(\sin^2\beta)}\frac{dV_{\text{eff}}}{dh_{1(2)}} = 0, \tag{2.46}$$

with the boundary conditions

$$\left.\frac{dh_{1,2}(\xi)}{d\xi}\right|_{\xi=0} = 0, \qquad h_{1,2}(\xi=\infty) = 0. \tag{2.47}$$

Actually, it is known that the bubble profile is approximated by a kink-type configuration

$$\rho_i(r) \sim v_i(T)\left[1 - \tanh\left(\frac{r-R}{L_w}\right)\right], \tag{2.48}$$

where R is the radius of the bubble and L_w is the wall width.

Finally, let us comment on the baryon number conservation condition at T_N. If the condition is satisfied at T_C, it is also satisfied at T_N. Since T_N is lower than T_C, and v_N, which is the VEV at T_N, is larger than v_C, it is found that v_N/T_N becomes larger than v_C/T_C. In addition, the sphaleron solution corresponds to a saddle point with the least energy in the path that connects two degenerate vacua, and it is the solution at finite temperature. Therefore, if the temperature is lower, the solution is displaced from the real solution. This situation results in the larger energy of the sphaleron, and it also causes the smaller value of $\zeta_{\text{sph}}(T)$.

2.3 Current Status of EWBG

The Standard Model was the first candidate for EWBG; however, it turned out that Sakharov's criteria are not satisfied in the framework. One remarkable thing is that discovery of the Higgs boson with 125 GeV mass [35, 36] excludes the possibility, and information related to the Higgs physics gives some restrictions on various new scenarios.

2.3.1 *The Standard Model EWBG*

EWBG in the SM cannot function due to the following two reasons:

1. Electroweak phase transition becomes crossover for $m_h > 73\,\text{GeV}$,[10]
2. The CP violation in the SM is too small to generate the observed BAU.

The first reason is the strongest one that the SM fails in EWBG. Lattice calculations show that the region for $m_h > 73\,\text{GeV}$ corresponds to crossover [37–41], whereas the discovered Higgs has 125 GeV mass. And, one another problem is that the CP violation in the SM is too small to produce the observed BAU [42–45].[11] Employing naive dimensional analysis, one can estimate the size of CP violation through the Jarlskog determinant

$$A_{\text{CP}} = (m_t^2 - m_c^2)(m_t^2 - m_u^2)(m_c^2 - m_u^2)(m_b^2 - m_s^2)(m_b^2 - m_d^2)(m_s^2 - m_d^2)J \tag{2.49}$$

with $J = \text{Im}(V_{ub}V_{cb}V_{ub}^*V_{cd}^*) \simeq s_{12}s_{23}s_{13}\sin\delta_{\text{KM}} \simeq 3\times 10^{-5}$. The dimensionless CP phase δ_{CP} with $T_C \sim 100\,\text{GeV}$ is roughly given by

$$\delta_{\text{CP}} \sim \frac{A_{\text{CP}}}{T_C^{12}} \sim 10^{-20}. \tag{2.50}$$

The above value is obviously too small to explain $n_B/n_\gamma \sim 10^{-10}$. Then, our present consensus is that the CP phase of the SM cannot create enough baryon numbers, which leads to the demand of new physics beyond the SM for the successful EWBG.

2.3.2 *Physics Beyond the Standard Model*

As a result of the drawbacks of the SM EWBG, the next possibility is new physics. Since new physics generally brings new CP phase, the second problem can be easily solved. However, such a new CP phase may induce the EDMs, so we should also consider the experimental constraints. In other words, the EDM measurements can examine the scenarios. Regarding the achievement of the first-order phase transition with the correct Higgs mass, introduction of new boson helps to create a barrier needed for the phase transition.

---

[10]Perturbative calculation shows smaller value of $m_h \lesssim 42\,\text{GeV}$ for the first-order phase transition [1].

[11]To be exact, this is because the size of the CP violation in the SM depends on the structure of the Cabibbo-Kobayashi-Maskawa matrix.

2.3.2.1 Phase Transition in New Physics

As seen in Fig. 2.5, the first-order EWPT is that the effective potential has two degenerate vacua at $T = T_C$. In order to see an origin of the barrier, we consider the finite-temperature one-loop effective potential in the SM with one constant background field φ. It consists of three parts:

$$V_{\text{eff}}(\varphi) = V_0(\varphi) + V_1(\varphi) + V_1(\varphi, T). \tag{2.51}$$

$V_0(\varphi)$ is the tree-level potential

$$V_0(\varphi) = -\frac{\mu^2}{2}\varphi^2 + \frac{\lambda}{4}\varphi^4, \tag{2.52}$$

and $V_1(\varphi)$ is the one-loop Coleman-Weinberg potential at zero temperature [30, 31]

$$V_1(\varphi) = \sum_i n_i \frac{m_i^4(\varphi)}{64\pi^2}\left(\ln\frac{m_i^2(\varphi)}{\mu^2} - c_i\right) \tag{2.53}$$

where $i = h,\ G^0,\ G^\pm,\ W,\ Z,\ t,\ b$. n_i is the degrees of freedom of the particle given by $n_{h,G^0} = 1$, $n_{G^\pm} = 2$, $n_Q = 2\cdot 3$, $n_Z = 3$, $n_t = n_b = -4N_C$, c_i is $3/2$ $(5/6)$ for scalars and fermions (gauge bosons), and $m_i(\varphi)$ is a field-dependent mass of the species i. The third term is given by[32]

$$V_1(\varphi, T) = \frac{T^4}{2\pi^2}\left[\sum_{i=\text{boson}} n_i I_B(a_i^2) + \sum_{i=\text{fermion}} n_i I_F(a_t^2)\right], \tag{2.54}$$

where

$$I_B(a^2) = \int_0^\infty dx x^2 \log\left(1 - e^{\sqrt{x^2+a^2}}\right), \tag{2.55}$$

$$I_F(a^2) = \int_0^\infty dx x^2 \log\left(1 + e^{\sqrt{x^2+a^2}}\right), \tag{2.56}$$

with $a_i^2 = m_i^2(\varphi)/T^2$. Taking a limit of $m_i^2(\varphi)/T^2 \ll 1$ called the high-temperature expansion, one obtains

$$I_B(a^2) = -\frac{\pi^4}{45} + \frac{\pi^2}{12}a^2 - \frac{\pi}{6}(a^2)^{3/2} - \frac{a^4}{32}\left(\log\frac{a^2}{\alpha_B} - \frac{3}{2}\right) + \mathcal{O}(a^6) + \cdots, \tag{2.57}$$

$$I_F(a^2) = \frac{7\pi^4}{360} - \frac{\pi^2}{24}a^2 - \frac{a^4}{32}\left(\log\frac{a^2}{\alpha_F} - \frac{3}{2}\right) + \mathcal{O}(a^6) + \cdots, \tag{2.58}$$

with $\log \alpha_B = 2\log 4\pi + 3/2 - 2\gamma_E$ and $\log \alpha_F = 2\log \pi + 3/2 - 2\gamma_E$, and γ_E is the Euler constant. It is seen that only bosons bring negative contributions to the potential as $V_1(\varphi, T) \ni -\sum_{\text{boson}} \left[m_i^2(\varphi)\right]^{3/2} T$. Finally, the high-temperature expansion results in the following simple effective potential

$$V_{\text{eff}}^{\text{HE}} = D(T^2 - T_0^2)\varphi^2 - ET\varphi^3 + \frac{\lambda_T}{4}\varphi^4, \tag{2.59}$$

where

$$D = \frac{2m_W^2 + m_Z^2 + 2 + 2m_t^2}{8v^2}, \tag{2.60}$$

$$T_0^2 = \frac{m_h^2 - 8Bv^2}{4D}, \tag{2.61}$$

$$E = \frac{2m_W^3 + m_Z^3}{4\pi v^3}, \tag{2.62}$$

$$B = \frac{3}{64\pi^2 v^4}(2m_W^4 + m_Z^4 - 4m_t^4), \tag{2.63}$$

$$\lambda_T = \lambda - (\text{log corrections}). \tag{2.64}$$

With the potential at $T = T_C$, the first-order phase transition is described by

$$V_{\text{eff}}^{\text{HE}} = \frac{\lambda_T}{4}\varphi^2(\varphi - v_C)^2, \qquad v_C = \frac{2ET_C}{\lambda_{T_C}}, \tag{2.65}$$

where two degenerate vacua at $\varphi = 0$ and v_C are given. And, this expression leads to

$$\frac{v_C}{T_C} = \frac{2E}{\lambda_{T_C}}. \tag{2.66}$$

As we discussed earlier, the size of v_C/T_C is important for the baryon number preservation condition which decides whether created baryon asymmetry is left until the present Universe. The high-temperature expansion tells us that the ratio corresponds to that of the third and fourth terms of the effective potential. The third term plays a role in negative effect to the effective potential, and it comes from only the boson contribution. If we introduce new boson, the contribution is added to the coefficient of E.[12] Therefore, new physics with additional boson has a possibility of successful first-order electroweak phase transition.[13]

Currently, various new physics are studied in terms of phase transition. Here, we introduce two major scenarios for the first-order phase transitions:

1. Thermal loop-driven case.
2. Tree potential-driven case.

The first case is that thermal loop effect brings a cubic term of $-T(\varphi^2)^{3/2}$ which create a barrier in the effective potential. This case is applied to the SM, MSSM [47–55], and two Higgs doublet model [14, 56–63]. In this case, so-called non-decoupling effect is necessary for the first-order EWPT. It implies that the mass of new boson described by $m^2(\varphi) = g^2\varphi^2 + M^2$, where M and g are mass parameter and dimensionless coupling in the Lagrangian, receives dominant contribution from $g\varphi$. The cubic term is described by $-(m^2(\varphi))^{3/2}T \sim -M^3T$ if $M > g\varphi$, which results in no negative contribution to the effective potential. In other words, situation without dimensionful parameter easily tends to cause the first-order EWPT [14, 56]. However, such a non-decoupling limit needs large value of the dimensionless coupling, and it follows that the cutoff scale at which $|g| > 4\pi$ is close to the electroweak scale. For example, the cutoff scale is at 6.3 TeV in scale-invariant two Higgs doublet model [14] where dimensionful parameters are forbidden at tree level and electroweak symmetry breaking is radiatively caused.

The second case is that tree-level potential creates the barrier, and model with a real singlet is classified into this case [13, 64–82]. Here, a nonzero coupling between real singlet and $SU(2)$ Higgs is necessary for getting different phase transition from the SM case. Such a coupling can deviate the Higgs coupling from the SM value due to the mixing, and it becomes one of the signature of the phase transition.

In addition to the above two patterns, the barrier is also able to be caused by other scenarios [83]. For example, if we consider that dimension six operator of $(H^\dagger H)^3$ appears after heavy particle is integrated out at a new scale Λ, it brings a $-\varphi^4$ term to the potential [84–86].[14] Also, triplet extensions are proposed to fulfill the first-order phase transition [87–89].

A fascination for EWBG is high testability. In addition to the SM, the scenario in MSSM is presently excluded [53, 90–92]. In MSSM, right-handed stop is required to be $m_{\tilde{t}_R} < 120\,\text{GeV}$, and the scenario predicts that $\sigma(gg \to H \to VV)/\sigma(gg \to H \to VV)_{\text{SM}} \sim (2-3)$. However, the value found to be inconsistent with the LHC data, and the region for the light stop mass is also excluded by the direct search

[12] In principle, smaller value of λ is also able to cause larger v_C/T_C. However, it would be less possibility that the coupling λ is deviated from the SM value unless some peculiar parameters are chosen.

[13] It is also proposed that new fermion which strongly couples to the Higgs can strengthen the first-order phase transition [46], where $O(1)$ Yukawa coupling is needed.

[14] It turned out that the scale is somewhat low $\Lambda \sim 800\,\text{GeV}$.

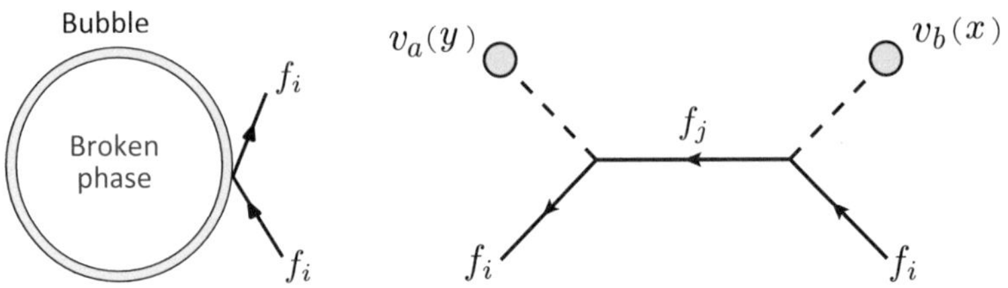

Fig. 2.11 Right: Interaction between fermion f_i and bubble walls indicated by v_a and v_b with $a, b = 1, 2$. Left: One example of particle diffusion caused by the process in the right diagram

at the LHC [53, 54, 90, 93]. What we can guess from the situation is that colored particles would not play a role in the first-order EWPT.

2.3.2.2 CP Violation in New Physics

CP-violating process should be involved in an interaction between particle and bubble wall. Figure 2.11 shows an example of the interaction between fermion f_i and the bubble wall represented by two different VEVs $v_{a,b}$ $(a, b = 1, 2)$. In the right figure, the CP phases exist in each vertices, and the diagram causes diffusions such as the left figure. This kind of CP-violating interactions is seen in several new physics such as next-to-MSSM [94, 95] and 2HDM [62, 96–100], and they can induce nonzero EDMs. For example, in the framework of the general 2HDM [100], the fermion can be the top quark with the complex Yukawa coupling, and the coupling produces the electron and quark EDMs at two-loop level. Moreover, if we assume that flavor-changing process is also able to supply the relevant CP violation to the BAU, it leads to the flavor-changing neutral current in broken phase. In [100], the top quark decays into the charm, and Higgs is discussed with relating EWBG, and the predicted value is $\mathrm{Br}(t \to ch) \simeq 0.15\%$ with one benchmark point. The process has been searched for by ATLAS and CMS, and the obtained limits are $\mathrm{Br}(t \to ch) < 0.46\%$ [101] and $\mathrm{Br}(t \to ch) < 0.40\%$ [102], respectively. Although the prediction is below the current limits, the search at HL-LHC [103] which aims to reach $\mathrm{Br}(t \to ch) < 0.015\%$ has a great deal of possibility to examine the scenario. Thus, not only the EDMs but also flavor physics open windows to look into EWBG in new physics.

At present, we do not know which new model is likely. Our understanding is that the EWBG hypothesis needs something new at electroweak scale. The possibility is able to be discussed in terms of an effective theory where heavy degrees of freedom are integrated out. So, in the next section, we would like to consider a situation in which only relevant electroweak-interacting particles exist.

References

1. A. Riotto, [hep-ph/9807454]
2. G. 't Hooft, Phys. Rev. D **14**, 3432 (1976) [Phys. Rev. D **18**, 2199 (1978)]
3. F.R. Klinkhamer, N.S. Manton, Phys. Rev. D **30**, 2212 (1984)
4. C.S. Lim, *Violation of CP Symmetry* (in Japanese) published by Saiensu-sha Co., Ltd. Publishers
5. N.H. Christ, Phys. Rev. D **21**, 1591 (1980)
6. J. Ambjorn, J. Greensite, C. Peterson, Nucl. Phys. B **221**, 381 (1983)
7. F.R. Klinkhamer, C. Rupp, J. Math. Phys. **44**, 3619 (2003) [hep-th/0304167]
8. Based on K. Funakubo's lecture slides held in the intensive course at Nagoya University on 20–21 June 2013
9. N.S. Manton, Phys. Rev. D **28**, 2019 (1983)
10. K. Funakubo, A. Kakuto, S. Tao, F. Toyoda, Prog. Theor. Phys. **114**, 1069 (2006) [hep-ph/0506156]
11. K. Funakubo, K. Fuyuto, E. Senaha, [arXiv:1612.05431 [hep-ph]]
12. K. Funakubo, E. Senaha, Phys. Rev. D **79**, 115024 (2009) [arXiv:0905.2022 [hep-ph]]
13. K. Fuyuto, E. Senaha, Phys. Rev. D **90**(1), 015015 (2014) [arXiv:1406.0433 [hep-ph]]
14. K. Fuyuto, E. Senaha, Phys. Lett. B **747**, 152 (2015)
15. V.A. Rubakov, M.E. Shaposhnikov, Usp. Fiz. Nauk **166**, 493 (1996) [Phys. Usp. **39**, 461 (1996)] [hep-ph/9603208]
16. A. Ringwald, Nucl. Phys. B **330**, 1 (1990)
17. O. Espinosa, Nucl. Phys. B **343**, 310 (1990)
18. D.T. Son, V.A. Rubakov, Nucl. Phys. B **422**, 195 (1994) [hep-ph/9310240]
19. S.H.H. Tye, S.S.C. Wong, Phys. Rev. D **92**(4), 045005 (2015) [arXiv:1505.03690 [hep-th]]
20. K. Funakubo, S.Otsuki, K. Takenaga, F. Toyoda, Prog. Theor. Phys. **87**, 663 (1992)
21. K. Funakubo, S. Otsuki, K. Takenaga, F. Toyoda, Prog. Theor. Phys. **89**, 881 (1993) [hep-ph/9211212]
22. P.B. Arnold, L.D. McLerran, Phys. Rev. D **36**, 581 (1987)
23. J. Ambjorn, T. Askgaard, H. Porter, M.E.Shaposhnikov, Nucl. Phys. B **353**, 346 (1991)
24. G.D. Moore, K. Rummukainen, Phys. Rev. D **61**, 105008 (2000) [hep-ph/9906259]
25. K. Funakubo, Prog. Theor. Phys. **96**, 475 (1996) [hep-ph/9608358]
26. N. Turok, Phys. Rev. Lett. **68**, 1803 (1992)
27. B.H. Liu, L.D. McLerran, N. Turok, Phys. Rev. D **46**, 2668 (1992)
28. L. Carson, X. Li, L.D. McLerran, R.T. Wang, Phys. Rev. D **42**, 2127 (1990)
29. G.D. Moore, Phys. Rev. D **59**, 014503 (1999) [hep-ph/9805264]
30. S.R. Coleman, E.J. Weinberg, Phys. Rev. D **7**, 1888 (1973)
31. R.Jackiw, Phys. Rev. D **9**, 1686 (1974)
32. L. Dolan, R. Jackiw, Phys. Rev. D **9**, 3320 (1974)
33. M.E. Carrington, J.I. Kapusta, Phys. Rev. D **47**, 5304 (1993)
34. A.D. Linde, Nucl. Phys. B **216**, 421 (1983); Erratum: [Nucl. Phys. B **223**, 544 (1983)]
35. G. Aad et al., [ATLAS Collaboration], Phys. Lett. B **716**, 1 (2013) [arXiv:1207.7214 [hep-ex]]
36. S. Chatrchyan et al., [CMS Collaboration], Phys. Lett. B **716**, 30 (2012) [arXiv:1207.7235 [hep-ex]]
37. K. Kajantie, M. Laine, K. Rummukainen, M.E. Shaposhnikov, Phys. Rev. Lett. **77**, 2887 (1996) [hep-ph/9605288]
38. K. Rummukainen, M. Tsypin, K. Kajantie, M. Laine, M.E. Shaposhnikov, Nucl. Phys. B **532**, 283 (1998) [hep-lat/9805013]
39. F. Csikor, Z. Fodor, J. Heitger, Phys. Rev. Lett. **82**, 21 (1999) [hep-ph/9809291]
40. Y. Aoki, F. Csikor, Z. Fodor, A. Ukawa, Phys. Rev. D **60**, 013001 (1999) [hep-lat/9901021]
41. M. Laine, K. Rummukainen, Nucl. Phys. Proc. Suppl. **73**, 180 (1999) [hep-lat/9809045]
42. M.B. Gavela, P. Hernandez, J. Orloff, O. Pene, Mod. Phys. Lett. A **9**, 795 (1994) [hep-ph/9312215]

43. M.B. Gavela, P. Hernandez, J. Orloff, O. Pene, C. Quimbay, Nucl. Phys. B **430**, 382 (1994) [hep-ph/9406289]
44. P. Huet, E. Sather, Phys. Rev. D **51**, 379 (1995) [hep-ph/9404302]
45. T. Konstandin, T. Prokopec, M.G. Schmidt, Nucl. Phys. B **679**, 246 (2004) [hep-ph/0309291]
46. M. Carena, A. Megevand, M. Quiros, C.E.M. Wagner, Nucl. Phys. B **716**, 319 (2005) [hep-ph/0410352]
47. A. Brignole, J.R. Espinosa, M. Quiros, F. Zwirner, Phys. Lett. B **324**, 181 (1994) [hep-ph/9312296]
48. J.R. Espinosa, M. Quiros, F. Zwirner, Phys. Lett. B **307** (1993) 106 [hep-ph/9303317]
49. M. Carena, M. Quiros, C.E.M. Wagner, Phys. Lett. B **380**, 81 (1996) [hep-ph/9603420]
50. D. Delepine, J.M. Gerard, R. Gonzalez Felipe, J. Weyers, Phys. Lett. B **386**, 183 (1996) [hep-ph/9604440]
51. J.M. Cline, K. Kainulainen, Nucl. Phys. B **482**, 73 (1996) [hep-ph/9605235]
52. M. Laine, K. Rummukainen, Nucl. Phys. B **535**, 423 (1998) [hep-lat/9804019]
53. T. Cohen, D.E. Morrissey, A. Pierce, Phys. Rev. D **86**, 013009 (2012) [arXiv:1203.2924 [hep-ph]]
54. M. Carena, G. Nardini, M. Quiros, C.E.M. Wagner, JHEP **1302**, 001 (2013) [arXiv:1207.6330 [hep-ph]]
55. M. Carena, G. Nardini, M. Quiros, C.E.M. Wagner, Nucl. Phys. B **812**, 243 (2009) [arXiv:0809.3760 [hep-ph]]
56. K. Funakubo, A. Kakuto, K. Takenaga, Prog. Theor. Phys. **91**, 341 (1994) [hep-ph/9310267]
57. A.T. Davies, C.D. froggatt, G. Jenkins, R.G. Moorhouse, Phys. Lett. B **336**, 464 (1994)
58. J.M. Cline, P.A. Lemieux, Phys. Rev. D **55**, 3873 (1997) [hep-ph/9609240]
59. S. Kanemura, Y. Okada, E. Senaha, Phys. Lett. B **606**, 361 (2005)
60. L. Fromme, S.J. Huber, M. Seniuch, JHEP **0611**, 038 (2006) [hep-ph/0605242]
61. G.C. Dorsch, S.J. Huber, J.M. No, JHEP **1310**, 029 (2013) [arXiv:1305.6610 [hep-ph]]
62. C.W. Chiang, K. Fuyuto, E. Senaha, Phys. Lett. B **762**, 315 (2016) [arXiv:1607.07316 [hep-ph]]
63. G.C. Dorsch, S.J. Huber, K. Mimasu, J.M. No, [arXiv:1705.09186 [hep-ph]]
64. J.R. Espinosa, M. Quiros, Phys. Lett. B **305**, 98 (1993) [hep-ph/9301285]
65. J. Choi, R.R. Volkas, Phys. Lett. B **317**, 385 (1993)
66. K.E.C. Benson, Phys. Rev. D **48**, 2456 (1993)
67. G.C. Branco, D. Delepine, D. Emmanuel-Costa, F.R. Gonzalez, Phys. Lett. B **442**, 229 (1998) [hep-ph/9805302]
68. S.W. Ham, Y.S. Jeong, S.K. Oh, J. Phys. G **31**(8), 857 (2005) [hep-ph/0411352]
69. A. Ahriche, Phys. Rev. D **75**, 083522 (2007) [hep-ph/0701192]
70. J.R. Espinosa, M. Quiros, Phys. Rev. D **76**, 076004 (2007) [hep-ph/0701145]
71. A. Noble, M. Perelstein, Phys. Rev. D **78**, 063518 (2008) [arXiv:0711.3018 [hep-ph]]
72. J.R. Espinosa, T. Konstandin, J.M. No, M. Quiros, Phys. Rev. D **78**, 123528 (2008) [arXiv:0809.3215 [hep-ph]]
73. S. Das, P.J. Fox, A. Kumar, N. Weiner, JHEP **1011**, 108 (2010)
74. A. Ashoorioon, T. Konstandin, JHEP **0907**, 086 (2009) [arXiv:0904.0353 [hep-ph]]
75. D.J.H. Chung, A.J. Long, Phys. Rev. D **84**, 103513 (2011)
76. J.R. Espinosa, T. Konstandin, F. Riva, Nucl. Phys. B **854**, 592 (2012) [arXiv:1107.5441 [hep-ph]]
77. V. Barger, D.J.H. Chung, A.J. Long, L.T. Wang, Phys. Lett. B **710**, 1 (2012) [arXiv:1112.5460 [hep-ph]]
78. P.H. Damgaard, D. O'Connell, T.C. Petersen, A. Tranberg, Phys. Rev. Lett. **111**(22), 221804 (2013)
79. S. Profumo, M.J. Ramsey-Musolf, G. Shaughnessy, JHEP **0708**, 010 (2007)
80. S. Profumo, M.J. Ramsey-Musolf, C.L. Wainwrightm, P. Winslow, Phys. Rev. D **91**(3), 035018 (2015)
81. J.M. Cline, K. Kainulainen, JCAP **1301**, 012 (2013)
82. C.Y. Chen, J. Kozaczuk, I.M. Lewis, [arXiv:1704.05844 [hep-ph]]

83. D.J.H. Chung, A.J. Long, L.T. Wang, Phys. Rev. D **87**(2), 023509 (2013) [arXiv:1209.1819 [hep-ph]]
84. C. Grojean, G. Servant, J.D. Wells, Phys. Rev. D **71**, 036001 (2005) [hep-ph/0407019]
85. B. Grinstein, M. Trott, Phys. Rev. D **78**, 075022 (2008) [arXiv:0806.1971 [hep-ph]]
86. C. Delaunay, C. Grojean, J.D. Wells, JHEP **0804**, 029 (2008) [arXiv:0711.2511 [hep-ph]]
87. H.H. Patel, M.J. Ramsey-Musolf, Phys. Rev. D **88**, 035013 (2013) [arXiv:1212.5652 [hep-ph]]
88. N. Blinov, J. Kozaczuk, D.E. Morrissey, C. Tamarit, Phys. Rev. D **92**(3), 035012 (2015) [arXiv:1504.05195 [hep-ph]]
89. S. Inoue, G. Ovanesyan, M.J. Ramsey-Musolf, Phys. Rev. D **93**, 015013 (2016) [arXiv:1508.05404 [hep-ph]]
90. D. Curtin, P. Jaiswal, P. Meade, JHEP **1208**, 005 (2012)
91. A. Katz, M. Perelstein, JHEP **1407**, 108 (2014) [arXiv:1401.1827 [hep-ph]]
92. A. Katz, M. Perelstein, M.J. Ramsey-Musolf, P. Winslow, Phys. Rev. D **92**(9), 095019 (2015) [arXiv:1509.02934 [hep-ph]]
93. K. Krizka, A. Kumar, D.E. Morrissey, Phys. Rev. D **87**(9), 095016 (2013)
94. K. Cheung, T.J. Hou, J.S. Lee, E. Senaha, Phys. Lett. B **710**, 188 (2012) [arXiv:1201.3781 [hep-ph]]
95. J. Kozaczuk, S. Profumo, C.L. Wainwright, Phys. Rev. D **87**(7), 075011 (2013) [arXiv:1302.4781 [hep-ph]]
96. S. Tulin, P. Winslow, Phys. Rev. D **84**, 034013 (2011) [arXiv:1105.2848 [hep-ph]]
97. T. Liu, M.J. Ramsey-Musolf, J. Shu, Phys. Rev. Lett. **108**, 221301 (2012) [arXiv:1109.4145 [hep-ph]]
98. J. Shu, Y. Zhang, Phys. Rev. Lett. **111**(9), 091801 (2013) [arXiv:1304.0773 [hep-ph]]
99. H.K. Guo, Y.Y. Li, T. Liu, M. Ramsey-Musolf, J. Shu, Phys. Rev. D **96**(11), 115034 (2017) [arXiv:1609.09849 [hep-ph]]
100. K. Fuyuto, W.S. Hou, E. Senaha, Phys. Lett. B **776**, 402 (2018) [arXiv:1705.05034 [hep-ph]]
101. G. Aad et al., [ATLAS Collaboration], JHEP **1512**, 061 (2015) [arXiv:1509.06047 [hep-ex]]
102. V. Khachatryan et al., [CMS Collaboration], JHEP **1702**, 079 (2017) [arXiv:1610.04857 [hep-ex]]
103. ATLAS Collaboration, ATL-PHYS-PUB-2013-012

Chapter 3
Model

Abstract An extension of the bosonic sector and CP phase are required to construct successful EWBG. The former plays a role in the negative contribution to the effective potential of the Higgs, and the latter is essential for generating the observed BAU. As seen in the light stop scenario, the introduction of the new colored particle into the model would be disfavored by the current LHC results. Regarding the CP phase, it should appear in the electroweak interaction with the Higgs VEVs. Taking into account this situation, the possible setup is as follows:

$$\text{UV-complete models} \supset \text{Extended Higgs sectors} + \text{EW-interacting fermions},$$

where it is assumed that irrelevant heavy particles are integrated out whatever a UV-complete model is. In this chapter, we focus on the extended model introducing new particles: two Higgs doublets, a real singlet, and EW-interacting fermions.

Keywords Real singlet · 1st order electroweak phase transition

3.1 The Model

Particle contents are shown in Table 3.1. This kind of setup applies to next-to-MSSM [1, 2], bino-driven EWBG in MSSM [3], and $U(1)'$-MSSM [4]. New scalars and fermions have the same charges under the SM interactions. For scalar sectors, the Lagrangian is given by

$$\mathcal{L}_{\text{scalar}} = \sum_{i=1,2} \left(D_\mu \Phi_i\right)^\dagger D^\mu \Phi_i + \frac{1}{2}\partial_\mu S \partial^\mu S - V_{\text{2HDM}} - V_S - V_{\Phi S}, \tag{3.1}$$

This chapter is based on the following two articles: Kaori Fuyuto and Eibun Senaha, Phys. Rev. D **90**, 015015, Copyright ©2014 American Physical Society, Kaori Fuyuto, Junji Hisano and Eibun Senaha, Phys. Lett. B **755**, 491 (2016), Copyright ©2016 The Authors. Published by Elsevier B.V.

© Springer Nature Singapore Pte Ltd. 2018
K. Fuyuto, *Electroweak Baryogenesis and Its Phenomenology*, Springer Theses,
https://doi.org/10.1007/978-981-13-1008-9_3

Table 3.1 Particle contents in our model. Upper: Scalar contents. Lower: Fermion contents

Scalars	$SU(3)_C \times SU(2)_L \times U(1)_Y$	Z_2
Φ_1	$(\mathbf{1},\ \mathbf{2},\ 1/2)$	$-$
Φ_2	$(\mathbf{1},\ \mathbf{2},\ 1/2)$	$+$
S	$(\mathbf{1},\ \mathbf{1},\ 0)$	$-$
Fermions	$SU(3)_C \times SU(2)_L \times U(1)_Y$	Z_2
$\tilde{\Phi}_1$	$(\mathbf{1},\ \mathbf{2},\ 1/2)$	$-$
$\tilde{\Phi}_2$	$(\mathbf{1},\ \mathbf{2},\ 1/2)$	$+$
$\tilde{S}^0$	$(\mathbf{1},\ \mathbf{1},\ 0)$	$\pm$

where $\Phi_{1,2}$ is $SU(2)$ doublet, S is real singlet, and $D_\mu = \partial_\mu - ig_2\frac{\tau^a}{2}A^a_\mu - ig_1 Y B_\mu$. After the scalars get VEVs, they are cast into the form

$$\Phi_1(x) = \begin{pmatrix} \phi_1^+ \\ \phi_1^0 \end{pmatrix} = \begin{pmatrix} \phi_1^+ \\ \frac{1}{\sqrt{2}}\left(v_1 + h_1(x) + ia_1(x)\right) \end{pmatrix},$$

$$\Phi_2(x) = \begin{pmatrix} \phi_2^+ \\ \phi_2^0 \end{pmatrix} = \begin{pmatrix} \phi_2^+ \\ \frac{1}{\sqrt{2}}\left(v_1 + h_2(x) + ia_2(x)\right) \end{pmatrix}, \tag{3.2}$$

$$S(x) = v_S + h_S(x). \tag{3.3}$$

The potentials are given by

$$V_{2\text{HDM}} = m_1^2\Phi_1^\dagger\Phi_1 + m_2^2\Phi_2^\dagger\Phi_2 - \left(m_3^2\Phi_1^\dagger\Phi_2 + \text{h.c}\right) + \frac{\lambda_1}{2}(\Phi_1^\dagger\Phi_1)^2 + \frac{\lambda_2}{2}(\Phi_2^\dagger\Phi_2)^2$$
$$+ \lambda_3(\Phi_1^\dagger\Phi_1)(\Phi_2^\dagger\Phi_2) + \lambda_4(\Phi_1^\dagger\Phi_2)(\Phi_2^\dagger\Phi_1) + \left[\frac{\lambda_5}{2}(\Phi_1^\dagger\Phi_2)^2 + \text{h.c}\right], \tag{3.4}$$

$$V_S = \mu_S^3 S + \frac{m_S^2}{2}S^2 + \frac{\mu_S'}{3}S^3 + \frac{\lambda_S}{8}S^4, \tag{3.5}$$

$$V_{\Phi S} = \frac{\delta_1}{2}\Phi_1^\dagger\Phi_1 S^2 + \frac{\delta_2}{2}\Phi_2^\dagger\Phi_2 S^2 + \mu_1\Phi_1^\dagger\Phi_1 S + \mu_2\Phi_2^\dagger\Phi_2 S - \left(\mu_3\Phi_1^\dagger\Phi_2 S + \text{h.c}\right). \tag{3.6}$$

In order to avoid Higgs-mediated flavor-changing neutral current (FCNC) processes at tree level, we impose a softly broken Z_2 symmetry. As in the case of MSSM, we also assume another matter parity under which the SM fermions are even and the EW-interacting fermions are odd. The matter parity forbids a lepton number violating term such as $\bar{\tilde{\Phi}}_1\Phi_1 l_R$.

For the Yukawa interactions, we consider the so-called Type II as follows:

$$-\mathcal{L}_Y = \bar{q}_L f_2^{(u)}(i\tau^2\Phi_2^*)u_R + \bar{q}_L f_1^{(d)}\Phi_1 d_R + \bar{l}L f_1^{(e)}\Phi_1 e_R + \text{h.c.} \tag{3.7}$$

Table 3.2 Assignments of Z_2 symmetry for new EW-interacting fermions

	$\tilde{\Phi}_1$	$\tilde{\Phi}_2$	$\tilde{S}^0$
Type-A	−	+	+
Type-B	−	+	−
Type-C	−	+	+
Type-D	−	+	−

In this type, the top and bottom masses are given by

$$m_t = \frac{y_t}{\sqrt{2}} v_2 = \frac{y_t}{\sqrt{2}} v \sin\beta, \qquad m_b = \frac{y_b}{\sqrt{2}} v_1 = \frac{y_b}{\sqrt{2}} \cos\beta, \tag{3.8}$$

with $\tan\beta = v_2/v_1$ and $v \simeq 246\,\text{GeV}$. In terms of an estimation of the BAU, it does not matter which Yukawa-type interaction is chosen since we include only a contribution from the top quark.

The new EW-interacting fermions are given by

$$\tilde{\Phi}_1(x) = \begin{pmatrix} \tilde{\phi}_1^0 \\ \tilde{\phi}_1^- \end{pmatrix}, \quad \tilde{\Phi}_2(x) = \begin{pmatrix} \tilde{\phi}_2^+ \\ \tilde{\phi}_2^0 \end{pmatrix}, \quad \tilde{S}(x). \tag{3.9}$$

where the fields are described by two components fields. The Lagrangian is expressed by

$$\mathcal{L}_{\text{fer}} = \sum_{i=1,2} \bar{\tilde{\Phi}}_i i\bar{\sigma}^\mu D_\mu \tilde{\Phi}_i + \bar{\tilde{S}} i\bar{\sigma}^\mu \partial_\mu \tilde{S} - V_{\tilde{\Phi}\tilde{S}}. \tag{3.10}$$

The potential $V_{\tilde{\Phi}\tilde{S}}$ has four relevant Z_2 assignments to the BAU as in Table 3.2,

$$\begin{aligned}
V^{\text{A}}_{\tilde{\Phi}\tilde{S}} &= c_{11}\tilde{\Phi}_1^a \varepsilon_{ab} \Phi_1^b \tilde{S}^0 + c_{22}\tilde{\Phi}_2^a \varepsilon_{ab} (i\tau_2\Phi_2^*)^b \tilde{S}^0 + (\mu + \lambda S)\tilde{\Phi}_1^a \varepsilon_{ab} \tilde{\Phi}_2^b + \frac{\mu_{\tilde{S}}}{2} \tilde{S}^0 \tilde{S}^0 + \text{h.c.}, \\
V^{\text{B}}_{\tilde{\Phi}\tilde{S}} &= c_{12}\tilde{\Phi}_1^a \varepsilon_{ab} \Phi_2^b \tilde{S}^0 + c_{21}\tilde{\Phi}_2^a \varepsilon_{ab} (i\tau_2\Phi_1^*)^b \tilde{S}^0 + (\mu + \lambda S)\tilde{\Phi}_1^a \varepsilon_{ab} \tilde{\Phi}_2^b + \frac{\mu_{\tilde{S}}}{2} \tilde{S}^0 \tilde{S}^0 + \text{h.c.}, \\
V^{\text{C}}_{\tilde{\Phi}\tilde{S}} &= c_{11}\tilde{\Phi}_1^a \varepsilon_{ab} \Phi_1^b \tilde{S}^0 + c_{22}\tilde{\Phi}_2^a \varepsilon_{ab} (i\tau_2\Phi_2^*)^b \tilde{S}^0 + \mu\tilde{\Phi}_1^a \varepsilon_{ab} \tilde{\Phi}_2^b + \frac{1}{2}(\mu_{\tilde{S}} + \kappa S) \tilde{S}^0 \tilde{S}^0 + \text{h.c.}, \\
V^{\text{D}}_{\tilde{\Phi}\tilde{S}} &= c_{12}\tilde{\Phi}_1^a \varepsilon_{ab} \Phi_2^b \tilde{S}^0 + c_{21}\tilde{\Phi}_2^a \varepsilon_{ab} (i\tau_2\Phi_1^*)^b \tilde{S}^0 + \mu\tilde{\Phi}_1^a \varepsilon_{ab} \tilde{\Phi}_2^b + \frac{1}{2}(\mu_{\tilde{S}} + \kappa S) \tilde{S}^0 \tilde{S}^0 + \text{h.c.},
\end{aligned} \tag{3.11}$$

where each of $\tilde{\Phi}_1$ and $\tilde{\Phi}_2$ couples to $\tilde{S}$.

Defining four components spinors as in Appendix A.1, one obtain relevant interactions between the EW-interacting fermions and scalars

$$\mathcal{L}_{\rm fer}^{\rm A(B)} \ni \Big\{ \overline{\tilde{H}^0} \left(c_L^{\tilde{H}^0 \tilde{S}} P_L \phi_{1(2)}^0 + c_R^{\tilde{H}^0 \tilde{S}} P_R \phi_{2(1)}^0 \right) \tilde{S} + \overline{\tilde{H}^+} \left(c_L^{\tilde{H}^\pm \tilde{S}} P_L \phi_{1(2)}^+ + c_R^{\tilde{H}^\pm \tilde{S}} P_R \phi_{2(1)}^+ \right) \tilde{S} + ({\rm h.c}) \Big\} - \overline{\tilde{H}^+} \left(g_{\bar{\tilde{H}} \tilde{H} S}^S + i\gamma_5 g_{\bar{\tilde{H}} \tilde{H} S}^P \right) \tilde{H}^+ h_S - \overline{\tilde{H}^0} \left(g_{\bar{\tilde{H}} \tilde{H} S}^S + i\gamma_5 g_{\bar{\tilde{H}} \tilde{H} S}^P \right) \tilde{H}^0 h_S, \tag{3.12}$$

and

$$\mathcal{L}_{\rm fer}^{\rm C(D)} \ni \Big\{ \overline{\tilde{H}^0} \left(c_L^{\tilde{H}^0 \tilde{S}} P_L \phi_{1(2)}^0 + c_R^{\tilde{H}^0 \tilde{S}} P_R \phi_{2(1)}^0 \right) \tilde{S} + \overline{\tilde{H}^+} \left(c_L^{\tilde{H}^\pm \tilde{S}} P_L \phi_{1(2)}^+ + c_R^{\tilde{H}^\pm \tilde{S}} P_R \phi_{2(1)}^+ \right) \tilde{S} + ({\rm h.c}) \Big\} - \frac{1}{2} \overline{\tilde{S}} \left(g_{\bar{\tilde{S}} \tilde{S} S}^S + i\gamma_5 g_{\bar{\tilde{S}} \tilde{S} S}^P \right) \tilde{S} h_S. \tag{3.13}$$

It should be noted that, while the charged fermions couple to the real singlet in type A and B, there are no such kinds of interactions in type C and D.

3.1.1 Real-Singlet Limit

Our model owns both 2HDM and singlet, and they can play a role in the first-order phase transition. As discussed in the last section of Chap. 1, the scenario in 2HDM needs non-decoupling limit which means that dimensionless couplings are somewhat large, and it follows that cutoff scale is close to electroweak scale. So, our study considers that a barrier in the Higgs potential is caused by the singlet contribution. In what follows, we take a real-singlet limit such that the scalar potentials are cast into the form [5]

$$V_0 = -\mu_H^2 H^\dagger H + \lambda_H (H^\dagger H)^2 + \mu_{HS} H^\dagger H S + \frac{\lambda_{HS}}{2} H^\dagger H S^2 + \mu_S^3 S + \frac{m_S^2}{2} S^2 + \frac{\mu_S'}{3} S^3 + \frac{\lambda_S}{4} S^4, \tag{3.14}$$

where H is the Higgs doublet of the $SU(2)$. After two scalar fields get the VEVs, they are given by

$$H(x) = \begin{pmatrix} G^+(x) \\ \frac{1}{\sqrt{2}} \left(v + h(x) + iG^0(x) \right) \end{pmatrix}, \qquad S(x) = v_S + h_S(x). \tag{3.15}$$

For the real-singlet limit, we firstly consider a rotation of the scalar states by an angle β

$$\begin{pmatrix} h_1 \\ h_2 \end{pmatrix} = R(\beta) \begin{pmatrix} h'_1 \\ h'_2 \end{pmatrix}, \quad \begin{pmatrix} a_1 \\ a_2 \end{pmatrix} = R(\beta) \begin{pmatrix} G^0 \\ A \end{pmatrix}, \quad \begin{pmatrix} \phi_1^\pm \\ \phi_2^\pm \end{pmatrix} = R(\beta) \begin{pmatrix} G^\pm \\ H^\pm \end{pmatrix}. \tag{3.16}$$

with $0 \le \beta \le \pi/2$ and

$$R(\beta) = \begin{pmatrix} \cos\beta & -\sin\beta \\ \sin\beta & \cos\beta \end{pmatrix}, \tag{3.17}$$

which implies that the CP-odd and charged scalars are diagonalized. The CP-even states are obtained by rotating $(h'_1,\ h'_2)$ with an angle $(\alpha - \beta)$

$$\begin{pmatrix} h'_1 \\ h'_2 \end{pmatrix} = R(\alpha - \beta) \begin{pmatrix} H \\ h \end{pmatrix}, \tag{3.18}$$

where α is denoted as the mixing angle for the diagonalization of h_1 and h_2 with $-\pi/2 \le \alpha \le 0$ and $m_H > m_h$. In addition, seeing the ration of $R(\beta)$ in terms of the basis of $(\Phi_1\ \Phi_2)$,

$$\begin{pmatrix} \Phi'_1 \\ \Phi'_2 \end{pmatrix} = \begin{pmatrix} \cos\beta & \sin\beta \\ -\sin\beta & \cos\beta \end{pmatrix} \begin{pmatrix} \Phi_1 \\ \Phi_2 \end{pmatrix}, \tag{3.19}$$

one obtain

$$\Phi'_1 = \begin{pmatrix} G^+ \\ \frac{1}{\sqrt{2}}(v + h'_1 + iG^0) \end{pmatrix}, \qquad \Phi'_2 = \begin{pmatrix} H^+ \\ \frac{1}{\sqrt{2}}(h'_2 + iA) \end{pmatrix}. \tag{3.20}$$

This is called he Georgi basis [6] in which only one Higgs doublet has the VEV. Note that taking $\sin(\beta - \alpha) = 1$ implies that $H = h'_2$ and $h = h'_1$. With the diagonalized masses, the dimensionless couplings λ_is are described by

$$\lambda_1 = \frac{1}{v^2 \cos^2\beta}\left(m_{H_1}^2 \sin^2\alpha + m_{H_2}^2 \cos^2\alpha - M^2 \sin^2\beta\right), \tag{3.21}$$

$$\lambda_2 = \frac{1}{v^2 \sin^2\beta}\left(m_{H_1}^2 \cos^2\alpha + m_{H_2}^2 \sin^2\alpha - M^2 \cos^2\beta\right), \tag{3.22}$$

$$\lambda_3 = \frac{\sin 2\alpha}{v^2 \sin 2\beta}(-m_{H_1}^2 + m_{H_1}^2) - \frac{1}{v^2}(M^2 - 2m_{H^\pm}^2), \tag{3.23}$$

$$\lambda_4 = \frac{1}{v^2}(M^2 + m_A^2 - 2m_{H^\pm}^2), \tag{3.24}$$

$$\lambda_5 = \frac{1}{v^2}(M^2 - m_A^2), \tag{3.25}$$

where $M^2 = m_3^2/\sin\beta\cos\beta$. Here, setting $\sin(\beta - \alpha) = \tan\beta = 1$ results in $\lambda_1 = \lambda_2$.

Next, assuming that $\mathrm{Im}(m_3^2) = 0$ and $\mathrm{Im}(\lambda_5) = 0$, we consider the tadpole conditions of the scalar potentials with respect to v_1 and v_2. m_1^2 and m_1^2 are described by

$$m_1^2 = m_3^2 \tan\beta - \frac{\lambda_1}{2} v^2 \cos^2\beta - \frac{1}{2}(\lambda_3 + \lambda_4 + \lambda_4) v^2 \sin^2\beta - \frac{\delta_1}{2} v_S^2 - \mu_1 v_S + \mu_3 v_S \tan\beta, \tag{3.26}$$

$$m_2^2 = m_3^2 \cot\beta - \frac{\lambda_2}{2} v^2 \sin^2\beta - \frac{1}{2}(\lambda_3 + \lambda_4 + \lambda_5) v^2 \cos^2\beta - \frac{\delta_2}{2} v_S^2 - \mu_2 v_S + \mu_3 v_S \cot\beta, \tag{3.27}$$

where $\lambda \equiv \lambda_1 = \lambda_2$. It is understood that $m_1^1 = m_2^2$ if $\tan\beta = 1,\ \delta_1 = \delta_2,\ \mu_1 = \mu_2$.

With the assumptions and definitions of $m_1^2 = m_2^2 \equiv m^2$ and $\lambda_1 = \lambda_2 \equiv \lambda$, the two Higgs doublet potential of V_{2HDM} is described by

$$\begin{aligned} V_{\mathrm{2HDM}} = m^2 \left(\Phi_1^\dagger \Phi_1 + \Phi_2^\dagger \Phi_2\right) - m_3^2 \left(\Phi_1^\dagger \Phi_2 + \text{h.c.}\right) + \frac{\lambda}{2} \left\{ \left(\Phi_1^\dagger \Phi_1\right)^2 + \left(\Phi_2^\dagger \Phi_2\right)^2 \right\} \\ + \lambda_3 (\Phi_1^\dagger \Phi_1)(\Phi_2^\dagger \Phi_2) + \lambda_4 (\Phi_1^\dagger \Phi_1)(\Phi_2^\dagger \Phi_1) + \frac{\lambda_5}{2} \left\{ \left(\Phi_1^\dagger \Phi_2\right)^2 + \text{h.c.} \right\}, \end{aligned} \tag{3.28}$$

Taking that $\Phi_{1(2)} \sim H/\sqrt{2}$, one can rewrite the potential as

$$V_{\mathrm{2HDM}} = -\mu^2 H^\dagger H + \lambda_{\mathrm{eff}} (H^\dagger H)^2, \tag{3.29}$$

with $\mu^2 = -m^2 + m_3^2$ and $\lambda_{\mathrm{eff}} = (\lambda + \lambda_3 + \lambda_4 + \lambda_5)/4$. Also, we take that $\delta_1 = \delta_2 \equiv \lambda_{HS}$, $\mu_1 = \mu_2 \equiv \mu_{HS}$ and $\mu_3 = 0$. This requirement produces

$$V_{\Phi S} = \frac{\lambda_{HS}}{2} H^\dagger H S^2 + \mu_{HS} H^\dagger H S. \tag{3.30}$$

Thus, V_{2HDM} and $V_{\Phi S}$ are finally translated into the first and second lines in Eq. (3.14). In this limit, relevant particles are the $SU(2)$ Higgs H and real singlet S for the phase transition.

3.1.2 *Real-Singlet Model*

In the real-singlet model, the mass matrix is diagonalized by

$$\mathcal{M}_{\mathrm{dia}}^2 = \begin{pmatrix} m_{H_1}^2 & 0 \\ 0 & m_{H_2}^2 \end{pmatrix}$$

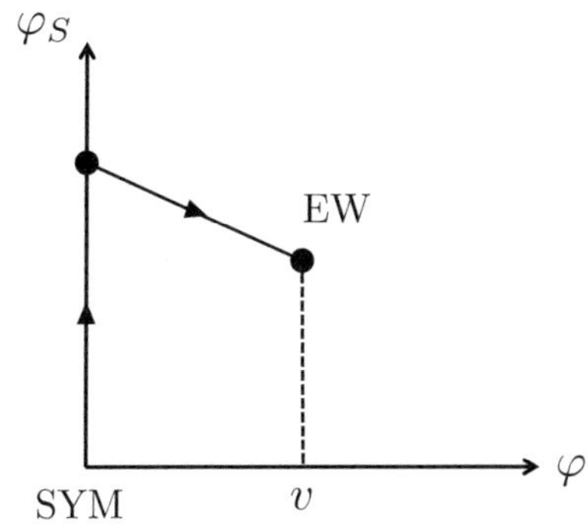

Fig. 3.1 Process of the phase transition. After symmetry breaking in the singlet direction occurs, that of the electroweak also starts

$$= R(\gamma)^T \begin{pmatrix} 2\lambda_H v^2 & \mu_{HS} v + \lambda_{HS} v v_S \\ \mu_{HS} v + \lambda_{HS} v v_S & -\frac{\mu_S^3}{v_S} + \mu_S' v_S + 2\lambda_S v_S^2 - \frac{\mu_{HS}}{2}\frac{v^2}{v_S} \end{pmatrix} R(\gamma), \tag{3.31}$$

with the physical states

$$\begin{pmatrix} H_1 \\ H_2 \end{pmatrix} = R(\gamma)^T \begin{pmatrix} h \\ h_S \end{pmatrix} = \begin{pmatrix} \cos\gamma & \sin\gamma \\ -\sin\gamma & \cos\gamma \end{pmatrix} \begin{pmatrix} h \\ h_S \end{pmatrix}, \tag{3.32}$$

where the tadpole conditions for φ and φ_S, which are constant background fields of h and h_S, are used in the mass matrix. Here, H_1 corresponds to the Higgs with the mass of 125 GeV, and H_2 is the new scalar.

Since the original interactions for h are scaled to $\cos\gamma$, the Higgs couplings to gauge bosons and fermions normalized to those in the SM are given by

$$\kappa_V = \frac{g_{H_1 VV}}{h_{hVV}^{\mathrm{SM}}} = \cos\gamma, \qquad \kappa_F = \frac{g_{H_1 ff}}{h_{hff}^{\mathrm{SM}}} = \cos\gamma. \tag{3.33}$$

The present experimental values are obtained by the ATLAS collaboration [7]

$$\kappa_V = 1.03 \pm 0.06, \qquad \kappa_F = 0.89^{+0.20}_{-0.15}, \tag{3.34}$$

and see also the current results by the CMS collaboration [8]. What we emphasize here is that the mixing is originated from the nonzero values of μ_{HS} and λ_{HS}, and they are also essential for obtaining different phase transition from the SM case.

As discussed in Ref. [9], one feature in the real-singlet model is two-step phase transitions. Figure 3.1 shows the phase transition, in which φ_S initially has a nonzero VEV with $\varphi = 0$, and then, φ starts to become the nonzero value. Let us parameterize φ and φ_S as

$$\varphi = r\cos\theta, \qquad \varphi_S = \varphi_S^0 + r\sin\theta, \tag{3.35}$$

where $-\pi/2 \leq \theta \leq 0$ and φ_S^0 is a minimum at $\varphi = 0$. In order to understand the structure of the phase transition analytically, we prepare the following simple potential:

$$V(\varphi, \varphi_S, T) = V_0(\varphi, \varphi_S) + \frac{1}{2}\Sigma_H(T)\varphi^2 + \frac{1}{2}\Sigma_S(T)\varphi_S^2 - ET\varphi^3, \tag{3.36}$$

where $\Sigma_H(T)$ and $\Sigma_S(T)$ are thermal masses for h and h_S [10]. The final term is denoted as the thermal effect from the finite-temperature 1-loop effective potential as we have seen in the Sect. 2.3. Using the parameterized fields, one rewrites the potential as

$$\begin{aligned} V(r, \theta, T) &= C_0 + C_1 r + C_2 r^2 - C_3 r^3 + C_4 r^4 \\ &= C_0 + C_1 r + C_4 r^2 \left[\left(r - \frac{C_3}{2C_4} \right)^2 - \frac{C_3^2}{4C_4^2} + \frac{C_2}{C_4} \right], \end{aligned} \tag{3.37}$$

where

$$\begin{aligned} C_0 &= \varphi_S^0 \left[\mu_S^3 + \frac{1}{2}\left(m_S^2 + \Sigma_S(T) \right)\varphi_S^0 + \frac{\mu_S'}{3}(\varphi_S^0)^2 + \frac{\lambda_S}{4}(\varphi_S^0)^3 \right], \\ C_1 &= \sin\beta \left[\mu_S^3 + (m_S^2 + \Sigma_S(T))\varphi_S^0 + \mu_S'(\varphi_S^0)^2 + (\varphi_S^0)^3\lambda_S \right], \\ C_2 &= \left[\frac{1}{2}\left(-\mu_H^2 + \Sigma_H(T) \right)\cos^2\theta + \frac{1}{2}\left(m_S^2 + \Sigma_S(T) \right)\sin^2\theta \right. \\ &\quad \left. + \varphi_S^0 \left(\frac{\mu_{HS}}{2}\cos^2\theta + \mu_S'\sin^2\theta \right) + \frac{1}{2}(\varphi_S^0)^2 \left(\frac{\lambda_{HS}}{2}\cos^2\theta + 3\lambda_S\sin^2\theta \right) \right], \\ C_3 &= ET\cos^3\theta - \frac{1}{2}\cos^2\theta\sin\theta\left(\mu_{HS} + \lambda_{HS}\varphi_S^0 \right) - \sin^3\theta\left(\frac{\mu'}{3} + \lambda_S\varphi_S^0 \right), \\ C_4 &= \frac{1}{4}\left(\lambda_H\cos^4\theta + \lambda_{HS}\cos^2\theta\sin^2\theta + \lambda_S\sin^4\theta \right). \end{aligned} \tag{3.38}$$

Here, the constant term is able to be dropped by shift in vacuum energy. It is required that $C_1 = 0$ and $C_2 = C_3^2/4C_4$ in order to get two degenerate minima at the critical temperature T_C, and the situation is described by

$$V(r_C, \theta_C, T_C) = C_4 r^2 (r - r_C)^2, \tag{3.39}$$

with $r_C = C_3(T_C)/2C_4$. Thus, we obtain

$$\begin{aligned} \frac{v_C}{T_C} &= \frac{r_C\cos\theta_C}{T_C} \\ &= \frac{E\cos^4\theta_C - \cos\theta_C\sin\theta_C\left[\left(\mu_{HS} + \lambda_{HS}\varphi_S^0 \right)\cos^2\theta_C/2 + \left(\mu_S'/3 + \lambda_S\varphi_S^0 \right)\sin^2\theta_C \right]/T_C}{\left(\lambda_H\cos^4\theta_C + \lambda_{HS}\cos^2\theta_C\sin^2\theta_C + \lambda_S\sin^4\theta_C \right)/2}. \end{aligned} \tag{3.40}$$

It is found that the strong phase transition is related to absolute value of the square bracket in the numerator.

3.2 Phase Transition in the Real-Singlet Model

The effective potential we employ is given by

$$V_{\rm eff}(\varphi,\ \varphi_S,\ T) = V_0(\varphi,\ \varphi_S) + V_1(\varphi, \varphi_S) + V_1(\varphi,\ \varphi_S,\ T) + V_{\rm daisy}(\varphi,\ \varphi_S,\ T), \tag{3.41}$$

where the tree-level potential

$$\begin{aligned} V_0(\varphi,\ \varphi_S) =& -\frac{\mu_H^2}{2}\varphi^2 + \frac{\lambda_H}{4}\varphi^4 + \frac{\mu_{HS}}{2}\varphi^2\varphi_S + \frac{\lambda_{HS}}{4}\varphi^2\varphi_S^2 \\ &+ \mu_S^3\varphi_S + \frac{m_S^2}{2}\varphi_S^2 + \frac{\mu_S'}{\varphi_S^3} + \frac{\lambda_S}{4}\varphi_S^4, \end{aligned} \tag{3.42}$$

the 1-loop Coleman-Weinberg potential

$$V_1(\varphi,\ \varphi_S) = \sum_i n_i \frac{\bar{m}_i^4(\varphi, \varphi_S)}{64\pi^2}\left(\ln\frac{\bar{m}_i^2(\varphi, \varphi_S)}{\mu^2} - c_i\right), \tag{3.43}$$

and the finite-temperature 1-loop potential

$$V_1(\varphi,\ \varphi_S,\ T) = \sum_i n_i \frac{T^4}{2\pi^2} I_{B,F}\left(\frac{\bar{m}_i^2(\varphi, \varphi_S)}{T^2}\right). \tag{3.44}$$

Here, in addition to the SM particles, a contribution from the real singlet is included. It should be noted that the zero-temperature 1-loop potential takes the Landau gauge ($\xi = 0$), and the problem is discussed in [11] with the $\hbar$ expansion. The last term is the so-called daisy contributions [12]

$$V_{\rm daisy}(\varphi,\ \varphi_S,\ T) = -\sum_{n_j} \frac{T}{12\pi}\left[\left\{\bar{M}_j^2(\varphi,\ \varphi_S,\ T)\right\}^{\frac{3}{2}} - \left\{\bar{m}_j^2(\varphi,\ \varphi_S,\ T)\right\}^{\frac{3}{2}}\right], \tag{3.45}$$

where the thermally corrected boson masses $\bar{M}_j^2(\varphi,\ \varphi_S,\ T) = \bar{m}_j^2(\varphi, \varphi_S) + \Pi_j(T)$ with the thermal masses $\Pi_j(T)$ [10, 12]. The daisy contributions correspond to resummations, which deals with breakdown of the perturbation at the high temperature [13]. As one example, let us briefly see a naive dimensional analysis in the ϕ^4 theory with describing all momentums and energy scales as the temperature T.

Figure 3.2 shows one and two loops in the ϕ^4 theory. Since the one-loop diagram has a quadratic divergence, it can be described by λT^2 with a dimensionless coupling λ. Upper loop in the two-loop diagram has the same structure as the one-loop diagram, while the lower loop is described by the first derivative of the 1-loop

Fig. 3.2 One- and two-loop diagrams in the ϕ^4 theory

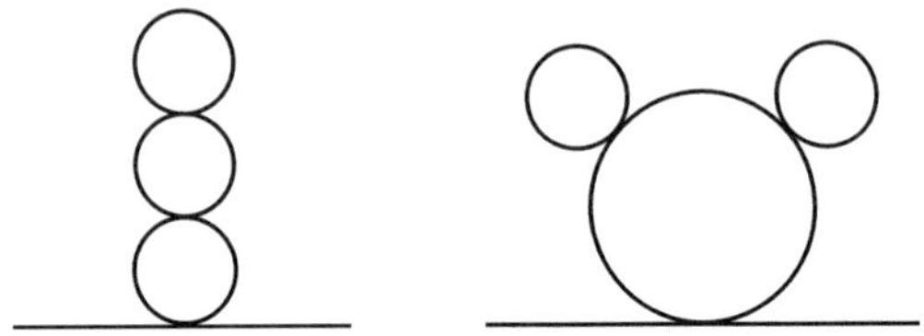

Fig. 3.3 Three-loop diagrams in the ϕ^4 theory

Fig. 3.4 Daisy diagram

diagram with respect to mass square m^2. The two-loop diagram finally leads to $(\lambda T^2)\ \lambda T/m$. Similarly, left cactus diagram in Fig. 3.3 produces $(\lambda T^2)\ (\lambda T/m)^2$, and right non-cactus diagram has $(\lambda T^2)^2\ \lambda T/m^3$. Taking ratios of these diagrams, one find that

$$\frac{\text{Non-cactus three loop}}{\text{Two loop}} \sim \frac{\lambda T^2}{m^2}, \tag{3.46}$$

$$\frac{\text{Non-cactus three loop}}{\text{Cactus loop}} \sim \frac{T}{m}. \tag{3.47}$$

It is understood that adding one-vertex bubble to the 1-loop diagram needs an extra factor of $\lambda T^2/m^2$, and the perturbation is breakdown if $\lambda T^2/m^2 \gtrsim 1$. The second relation implies that the diagram in which several one-vertex bubbles are attached to the central loop becomes dominant at high temperature. Thus, daisy diagram in Fig. 3.4 is non-negligible contribution, whose size is roughly given by

$$(\text{Daisy diagram}) \sim \frac{\lambda^2 T^3}{m} \left(\frac{\lambda T^2}{m^2}\right)^{n-1}. \tag{3.48}$$

This situation can also be regarded as an IR divergence of the zero mode in the boson propagator. Therefore, in order to deal with the problem, we add thermal effects to the boson masses such as $m^2(\varphi) \rightarrow m^2(\varphi) + \lambda T^2$, which is called resummation. Our treatment in Eq. (3.45) amounts to the resummation of only the zero-mode propagator. Recent study about the resummation is discussed in [14].

Using the potential in Eq. (3.41), we find v_C and T_C. Once the critical VEV and temperature are obtained, the sphaleron energy is numerically evaluated by solving the equations of motion in the real-singlet model. Here, we employ the following ansatz for the $SU(2)$ gauge, Higgs, and real singlet:

$$A_i(\mu, r, \theta, \phi) = -\frac{i}{g_2} f(r)\partial_i U(\mu, \theta, \phi)U^{-1}(\mu, \theta, \phi), \tag{3.49}$$

$$H(\mu, r, \theta, \phi) = \frac{v}{\sqrt{2}}\left[(1-h(r))\begin{pmatrix} 0 \\ e^{-i\mu}\cos\mu \end{pmatrix} + h(r)U(\mu, \theta, \phi)\begin{pmatrix} 0 \\ 1 \end{pmatrix}\right], \tag{3.50}$$

$$S(\mu, r, \theta, \phi) = v_S k(r), \tag{3.51}$$

where $U(\mu, \theta, \phi)$ is defined in Eq. (2.19). The equations of motion are given by

$$\frac{d^2 f}{d\xi^2} = \frac{2}{\xi^2} f(1-f)(1-2f) - \frac{1}{4}h^2(1-f), \tag{3.52}$$

$$\frac{d}{d\xi}\left(\xi^2 \frac{dh}{d\xi}\right) = 2h(1-f)^2 + \frac{\xi^2}{g_2^2}\frac{1}{v^4}\frac{\partial V_{\text{eff}}}{\partial h}, \tag{3.53}$$

$$\frac{d}{d\xi}\left(\xi^2 \frac{dk}{d\xi}\right) = \frac{\xi^2}{g_2^2}\frac{1}{v^2 v_S^2}\frac{\partial V_{\text{eff}}}{\partial k}, \tag{3.54}$$

with the boundary conditions

$$\lim_{\xi\to 0} f(\xi) = 0, \qquad \lim_{\xi\to 0} h(\xi) = 0, \qquad \lim_{\xi\to 0} k'(\xi) = 0, \tag{3.55}$$

$$\lim_{\xi\to\infty} f(\xi) = 1, \qquad \lim_{\xi\to\infty} h(\xi) = 1, \qquad \lim_{\xi\to\infty} k(\xi) = 1. \tag{3.56}$$

The sphaleron energy is

$$E_{\text{sph}} = \frac{4\pi v}{g_2}\int_0^\infty d\xi\left[4\left(\frac{df}{d\xi}\right)^2 + \frac{8}{\xi^2}(f-f^2)^2 + \frac{\xi^2}{2}\left(\frac{dh}{d\xi}\right)^2 + h^2(1-f)^2 \right.$$
$$\left. + \frac{\xi^2}{2}\frac{v_S^2}{v^2}\left(\frac{dk}{d\xi}\right)^2 + \frac{\xi^2}{g_2^2 v^4} V_{\text{eff}}(h, k, T)\right]. \tag{3.57}$$

With the obtained sphaleron energy, we clarify regions that satisfies the baryon number conservation criterion. Our analysis focuses on evaluating $\mathcal{E}(T_C)$ without the zero-mode factors, so the condition we estimate is described by

$$\frac{v(T_C)}{T_C} > \frac{g_2}{4\pi\mathcal{E}(T_C)}\left[42.97 - 2\log\left(\frac{T}{100\ \text{GeV}}\right)\right] \equiv \zeta_{\text{sph}}(T_C). \tag{3.58}$$

References

1. K. Cheung, T.J. Hou, J.S. Lee, E. Senaha, Phys. Lett. B **710**, 188 (2012). [arXiv: 1201.3781 [hep-ph]]
2. J. Kozaczuk, S. Profumo, C.L. Wainwright, Phys. Rev. D **87**(7), 075011 (2013). [arXiv: 1302.4781 [hep-ph]]
3. Y. Li, S. Profumo, M. Ramsey-Musolf, Phys. Lett. B **673**, 95 (2009). [arXiv: 0811.1987 [hep-ph]]
4. E. Senaha, Phys. Rev. D **88**(5), 055014 (2013). [arXiv: 1308.3389 [hep-ph]]
5. K. Fuyuto, E. Senaha, Phys. Rev. D **90**(1), 015015 (2014). [arXiv: 1406.0433 [hep-ph]]
6. H. Georgi, D.V. Nanopoulos, Phys. Lett. B **82**, 95 (1979)
7. ATLAS-CONF-2017-047, https://atlas.web.cern.ch/Atlas/GROUPS/PHYSICS/CONFNOTES/ATLAS-CONF-2017-047/
8. CMS PAS HIG-16-040, https://cds.cern.ch/record/2264515
9. K. Funakubo, S. Tao, F. Toyoda, Prog. Theor. Phys. **114**, 369 (2005). [hep-ph/0501052]
10. J.R. Espinosa, T. Konstandin, F. Riva, Nucl. Phys. B **854**, 592 (2012). [arXiv: 1107.5441 [hep-ph]]
11. H.H. Patel, M.J. Ramsey-Musolf, JHEP **1107**, 029 (2011). [arXiv: 1101.4665 [hep-ph]]
12. M.E. Carrington, Phys. Rev. D **45**, 2933 (1992)
13. L. Dolan, R. Jackiw, Phys. Rev. D **9**, 3320 (1974)
14. D. Curtin, P. Meade, H. Ramani, arXiv: 1612.00466 [hep-ph]

Chapter 4
Baryon Number

Abstract New physics generally contains additional CP phases, and they appear in the interactions between the new EW-interacting fermions and the two Higgs doublets in our extended model. In this chapter, we discuss how the CP-violating interactions lead to producing the baryon asymmetry based on closed time path formalism.

Keywords Baryon asymmetry · CP-violating interactions · Closed time path formalism

4.1 Derivation of the Diffusion Equations

Procedures for the estimation of the BAU are briefly the following:

1. Prepare diffusion equations for charges of the left- (Q) and right-handed (T) top quarks, and store the Higgs sectors (H) in the symmetric phase.
2. Derive the left-handed number density n_L by solving the coupled diffusion equations.
3. Transform n_L to the baryon number n_B.

Formally, the diffusion equation for the number density n_a takes the form

$$\partial_t n_a + \nabla \cdot \mathbf{j}_a = \Gamma_a + S_a^{\mathrm{CPV}}, \tag{4.1}$$

where Γ_a is chirality-changing rate and S_a^{CPV} represents CP-violating source term. The chirality-changing rate has an effect of reducing the final baryon number, and we include the top Yukawa, the top mass, and the strong sphaleron interactions as the chirality-changing processes. On another front, the final expression of n_B is linearly proportional to the CP-violating source term, which means the second condition of the Sakharov criteria.

© Springer Nature Singapore Pte Ltd. 2018

K. Fuyuto, *Electroweak Baryogenesis and Its Phenomenology*, Springer Theses,
https://doi.org/10.1007/978-981-13-1008-9_4

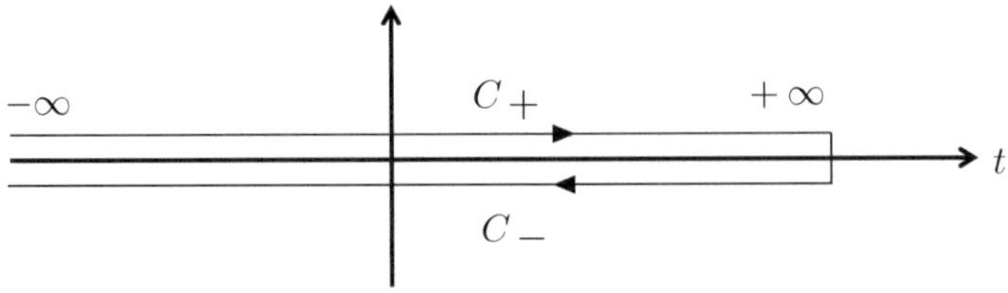

Fig. 4.1 The closed time path in the direction of t, which runs from $-\infty$ to $+\infty$ and then back to $-\infty$. Fields are distinguished by their placement on the first path of C_+ and the second path of C_-

In the evaluation of the baryon number, we use the closed time path (CTP) formalism [1–4], which is able to formulate Green's functions with incorporating the non-equilibrium phenomena. The closed time path implies that the time integration contour is closed running from $-\infty$ to $+\infty$ and back to $-\infty$ as in Fig. 4.1. The diffusion equations obtained by the CTP formalism are called the quantum Boltzmann equations (QBE's), and they can automatically take the CP-violating sources in. At present, the CTP formalism is regarded as the strongest tool for the calculation of the BAU based on the field theory.

Let us assume that the Hamiltonian of a system is described by a combination of H_0 and H_{int}, in which perturbative expansion is treated. An expectation value of an operator $\mathcal{O}_{\text{int}}$ at the time t is described by

$$\langle \mathcal{O} \rangle(t) \equiv \langle \psi_{\text{int}}(t)|\mathcal{O}_{\text{int}}|\psi_{\text{int}}(t)\rangle = \langle \psi_i|U^\dagger(t,t_i)\mathcal{O}_{\text{int}}U(t,t_i)|\psi_i\rangle, \tag{4.2}$$

with an initial state $|\psi_i\rangle$, and

$$U(t_2,t_1) = T\left[\exp\left\{-i\int_{t_1}^{t_2} dt' H_{\text{int}}(t')\right\}\right], \tag{4.3}$$

where T is the time-ordering operator. Since it is valid that $U(t_f,t_i)|\psi_i\rangle = e^{i\theta}|\psi_i\rangle$ under adiabatic assumption, the expectation value is written by

$$\begin{aligned}\langle \mathcal{O} \rangle(t) &= \langle \psi_i|U(t_i,t_f)U(t_f,t)\mathcal{O}_{\text{int}}U(t,t_i)|\psi_i\rangle \\ &= e^{-i\theta}\langle \psi_i|T\left[U(t_f,t_i)\mathcal{O}_{\text{int}}\right]|\psi_i\rangle \\ &= \frac{\langle \psi_i|T\left[U(t_f,t_i)\mathcal{O}_{\text{int}}\right]|\psi_i\rangle}{\langle \psi_i|U(t_f,t_i)\mathcal{O}_{\text{int}}|\psi_i\rangle}.\end{aligned} \tag{4.4}$$

At finite temperature, this expression is not correct since the expression of $U(t_f,t_i)|\psi_i\rangle = e^{i\theta}|\psi_i\rangle$ is not valid. The initial state is not the same as that of the final state anymore in a situation such as the early Universe where we cannot ignore cosmic expansion.[1] The situation we consider here is an area where a bubble expands and particles flow into it. Under these circumstances, the adiabatic evolution is no longer valid. Therefore, the above expectation value is given by

$$\langle \mathcal{O} \rangle(t) = \langle \psi_i | U(t_i, t_f) T \left[U(t_f, t_i) \mathcal{O}_{\text{int}} \right] | \psi_i \rangle. \tag{4.5}$$

Taking $t_i = -\infty$ and $t_f = +\infty$, one find that the route of t corresponds to the closed time path as in Fig. 4.1, and the expectation value is obtained without specifying the final states. It is also seen that the CTP formalism makes it possible to express quantum averages of operators at finite time including effects from the past collision. This is exactly the necessary method for evaluation of the BAU.

4.1.1 Green's Functions in the Framework of the CTP Formalism

Since there are positive and negative branches in the closed time path, propagator takes the 2×2 matrix form. For bosons, Green's function is defined as follows:

$$\hat{G}(x, y) \equiv \langle T_P \phi(x) \phi^\dagger(y) \rangle = \begin{pmatrix} G^t(x, y) & G^<(x, y) \\ G^>(x, y) & G^{\bar{t}}(x, y) \end{pmatrix} = \begin{pmatrix} ++ & +- \\ -+ & -- \end{pmatrix}, \tag{4.6}$$

where $+$ $(-)$ denotes the branch from $-\infty$ to $+\infty$ ($+\infty$ to $-\infty$) in the closed time path and

$$\begin{aligned} G^>(x, y) &= \langle \phi_-(x) \phi_+^\dagger(y) \rangle, \\ G^<(x, y) &= \langle \phi_-^\dagger(y) \phi_+(x) \rangle, \\ G^t(x, y) &= \left\langle T \left\{ \phi_+(x) \phi_+^\dagger(y) \right\} \right\rangle \\ &= \theta(x^0 - y^0) G^>(x, y) + \theta(y^0 - x^0) G^<(x, y), \\ G^{\bar{t}}(x, y) &= \left\langle \bar{T} \left\{ \phi_-(x) \phi_-^\dagger(y) \right\} \right\rangle \\ &= \theta(x^0 - y^0) G^<(x, y) + \theta(y^0 - x^0) G^>(x, y). \end{aligned} \tag{4.7}$$

In the same way, we define self-energies

$$\hat{\Sigma}(x, y) = \begin{pmatrix} \Sigma^t(x, y) & \Sigma^<(x, y) \\ \Sigma^>(x, y) & \Sigma^{\bar{t}}(x, y) \end{pmatrix}. \tag{4.8}$$

[1]This statement is also mentioned in Ref. [5].

The self-energies are related to Green's functions through the Schwinger-Dyson (SD) equations

$$\hat{G}(x,y) = \hat{G}^0(x,y) - i\int_C d^4z \int_C d^4w\, \hat{G}^0(x,z)\hat{\Sigma}(z,w)\hat{G}(w,y), \tag{4.9}$$

$$\hat{G}(x,y) = \hat{G}^0(x,y) - i\int_C d^4z \int_C d^4w\, \hat{G}(x,z)\hat{\Sigma}(z,w)\hat{G}^0(w,y), \tag{4.10}$$

where C denotes the closed time path and $\hat{G}^0(x,y)$ is the free propagator that follows

$$(\Box_x + m^2(x))\hat{G}^0(x,y) = -i\delta^{(4)}(x-y), \tag{4.11}$$

and $\delta_C^{(4)}(x-y)$ is defined by

$$\delta_C^{(4)}(x-y) = \delta^{(4)}(x-y)\tau^3 = \begin{cases} \delta^{(4)}(x-y) & \text{for } G^t(x,y), \\ -\delta^{(4)}(x-y) & \text{for } G^{\bar{t}}(x,y), \\ 0 & \text{for others.} \end{cases} \tag{4.12}$$

If $(\Box_x + m^2(x))$ acts on Eqs. (4.9) and (4.10), we obtain

$$(\Box_x + m^2(x))\hat{G}(x,y) = -i\delta^{(4)}(x-y)\tau^3 - \int_{-\infty}^{\infty} d^4w\, \hat{\Sigma}(x,w)\tau^3\hat{G}(w,y), \tag{4.13}$$

$$(\Box_y + m^2(x))\hat{G}(x,y) = -i\delta^{(4)}(x-y)\tau^3 - \int_{-\infty}^{\infty} d^4w\, \hat{G}(x,w)\tau^3\hat{\Sigma}(w,y). \tag{4.14}$$

This is because the second term on the right side comes from

$$\begin{aligned} &- i\int_C d^4z\, (\Box_x + m^2)\hat{G}^0(x,z)\hat{\Sigma}(z,w) \\ &= -i\left(\int_{-\infty}^{\infty} + \int_{+\infty}^{-\infty}\right) d^4z\, (\Box_x + m^2)\hat{G}^0(x,z)\hat{\Sigma}(z,w) \\ &= -\left\{\int_{-\infty}^{\infty} \delta^{(4)}(x-z)\begin{pmatrix}1&0\\0&0\end{pmatrix} - \int_{\infty}^{-\infty} \delta^{(4)}(x-z)\begin{pmatrix}0&0\\0&1\end{pmatrix}\right\} d^4z\hat{\Sigma}(z,w) \\ &= -\int_{-\infty}^{\infty} d^4z\, \delta^{(4)}(x-z)\mathbf{1}\hat{\Sigma}(z,w) \\ &= -\hat{\Sigma}(x,w), \end{aligned} \tag{4.15}$$

and then

$$\begin{aligned}
&-i\int_C d^4z\int_C d^4w\;(\Box_x+m^2)\hat{G}^0(x,z)\hat{\Sigma}(z,w)\hat{G}(w,y)\\
&=-\int_C d^4w\;\hat{\Sigma}(x,w)\hat{G}(w,y)\\
&=-\left(\int_{-\infty}^{\infty}+\int_{+\infty}^{-\infty}\right)d^4w\;[\hat{\Sigma}(x,w)]_{ij}[\hat{G}(w,y)]_{jk}\\
&=-\int_{-\infty}^{\infty}d^4w\left([\hat{\Sigma}(x,w)]_{i+}[\hat{G}(w,y)]_{+k}-[\hat{\Sigma}(x,w)]_{i-}[\hat{G}(w,y)]_{-k}\right)\\
&=-\int_{-\infty}^{\infty}d^4w\;\hat{\Sigma}(x,w)\tau^3\hat{G}(w,y). \qquad (4.16)
\end{aligned}$$

$\hat{G}_{11}(x,y)$ and $\hat{G}_{22}(x,y)$ in the above equations take forms

$$\begin{aligned}
(\Box_x+m^2(x))\hat{G}^t(x,y)=-i\delta^{(4)}(x-y)-\int_{-\infty}^{\infty}d^4w\;\Sigma^t(x,w)G^t(w,y)\\
-\Sigma^<(x,w)G^>(w,y), \qquad (4.17)
\end{aligned}$$

$$\begin{aligned}
(\Box_y+m^2(x))\hat{G}^t(x,y)=-i\delta^{(4)}(x-y)-\int_{-\infty}^{\infty}d^4w\;G^t(x,w)\Sigma^t(w,y)\\
-G^<(x,w)\Sigma^>(w,y), \qquad (4.18)
\end{aligned}$$

$$\begin{aligned}
(\Box_x+m^2(x))\hat{G}^{\bar{t}}(x,y)=i\delta^{(4)}(x-y)-\int_{-\infty}^{\infty}d^4w\;\Sigma^>(x,w)G^<(w,y)\\
-\Sigma^{\bar{t}}(x,w)G^{\bar{t}}(w,y), \qquad (4.19)
\end{aligned}$$

$$\begin{aligned}
(\Box_y+m^2(x))\hat{G}^{\bar{t}}(x,y)=i\delta^{(4)}(x-y)-\int_{-\infty}^{\infty}d^4w\;G^>(x,w)\Sigma^<(w,y)\\
-G^{\bar{t}}(x,w)\Sigma^{\bar{t}}(w,y). \qquad (4.20)
\end{aligned}$$

$\hat{G}_{12}(x,y)$ and $\hat{G}_{21}(x,y)$ become

$$(\Box_x+m^2(x))\hat{G}^<(x,y)=-\int_{-\infty}^{\infty}d^4w\;\Sigma^t(x,w)G^<(w,y)-\Sigma^<(x,w)G^{\bar{t}}(w,y), \qquad (4.21)$$

$$(\Box_y+m^2(x))\hat{G}^<(x,y)=-\int_{-\infty}^{\infty}d^4w\;G^t(x,w)\Sigma^<(w,y)-G^<(x,w)\Sigma^{\bar{t}}(w,y), \qquad (4.22)$$

$$(\Box_x + m^2(x))\hat{G}^>(x, y) = -\int_{-\infty}^{\infty} d^4w\ \Sigma^>(x, w)G^t(w, y) - \Sigma^{\bar{t}}(x, w)G^>(w, y), \tag{4.23}$$

$$(\Box_y + m^2(x))\hat{G}^>(x, y) = -\int_{-\infty}^{\infty} d^4w\ G^>(x, w)\Sigma^t(w, y) - G^{\bar{t}}(x, w)\Sigma^>(w, y). \tag{4.24}$$

Next, we write down the Noether current of the scalar field with Green's functions. The current of the scalar field is given by

$$J^\mu(x) = -iQ\left[\phi^\dagger(x)\{\partial^\mu\phi(x)\} - \{\partial^\mu\phi^\dagger(x)\}\phi(x)\right], \tag{4.25}$$

where Q is a conserved quantity that is associated with the transformation of $\phi \to \phi' = e^{i\alpha Q}\phi$.[2] This current can be expressed by $G^<(x, y)$

$$\begin{aligned} Q\ (\partial_x^\mu - \partial_y^\mu)G^<(x, y)\Big|_{y\to x} &= Q\ (\partial_x^\mu - \partial_y^\mu)\langle\phi_-^\dagger(y)\phi_+(x)\rangle\Big|_{y\to x} \\ &= Q\ \langle\phi_-^\dagger(y)\{\partial^\mu\phi_+(x)\} - \{\partial^\mu\phi_-^\dagger(y)\}\phi_+(x)\rangle\Big|_{y\to x} \\ &= i\langle J^\mu(x)\rangle. \end{aligned} \tag{4.26}$$

The derivative of the current is given by

$$\begin{aligned} \partial_{x\mu}\langle J^\mu(x)\rangle &= -iQ\partial_{x\mu}\ (\partial_x^\mu - \partial_y^\mu)G^<(x, y)\Big|_{y\to x} \\ &= -iQ\ (\Box_x - \Box_y)G^<(x, y)\big|_{y\to x}. \end{aligned}$$

In order to see the distribution of particles macroscopically, we take the center-of-mass coordinate system $X = \frac{x+y}{2}$ and $r = x - y$, which leads to

$$\begin{aligned} \frac{\partial n_\phi(X)}{\partial t_X} + \nabla_X \cdot \mathbf{j}_\phi(X) = iQ\int_{-\infty}^{\infty} d^4w\ \Big[&\Sigma^t(X, w)G^<(w, y) - \Sigma^<(x, w)G^{\bar{t}}(w, X) \\ &-G^t(X, w)\Sigma^<(w, X) + G^<(X, w)\Sigma^{\bar{t}}(w, X)\Big]. \end{aligned} \tag{4.27}$$

$\Sigma^{t,\bar{t}}$ and $G^{t,\bar{t}}$ are rewritten by

$$\Sigma^t(X, z) = \theta(X^0 - z^0)\Sigma^>(X, z) + \theta(z^0 - X^0)\Sigma^<(X, z), \tag{4.28}$$

$$\Sigma^{\bar{t}}(X, z) = \theta(z^0 - X^0)\Sigma^<(z, X) + \theta(X^0 - z^0)\Sigma^>(z, X), \tag{4.29}$$

[2]Here, the conserved quantity corresponds to a particle number.

$$G^t(X,z) = \theta(X^0 - z^0)G^>(X,z) + \theta(z^0 - X^0)G^<(X,z), \tag{4.30}$$

$$G^{\bar{t}}(X,z) = \theta(z^0 - X^0)G^<(z,X) + \theta(X^0 - z^0)G^>(z,X), \tag{4.31}$$

so Eq. (4.27) finally becomes

$$\begin{aligned}\frac{\partial n_\phi(X)}{\partial t_X} + \nabla_X \cdot \mathbf{j}_\phi(X) =& iQ\int_{-\infty}^{t_X} dz^0 \int_{-\infty}^{\infty} d^3\mathbf{z}\big[\Sigma^>(X,z)G^<(z,X) \\ & -\Sigma^<(X,z)G^>(z,X) - G^>(X,z)\Sigma^<(z,X) \\ & +G^<(X,Z)\Sigma^>(z,X)\big].\end{aligned} \tag{4.32}$$

This is exactly the diffusion equation for n_ϕ, and it expresses the time variation of the number density of n_ϕ. The time integration is performed from $-\infty$ to T, which implies that memory effect holding information about past history of the system appears.

Fermion parts are similarly extracted. The Green's function is

$$\hat{S}_{\alpha\beta}(x,y) \equiv \langle T_P \psi_\alpha(x)\bar{\psi}_\beta(y)\rangle = \begin{pmatrix} S^t_{\alpha\beta}(x,y) & S^<_{\alpha\beta}(x,y) \\ S^>_{\alpha\beta}(x,y) & S^{\bar{t}}_{\alpha\beta}(x,y)\end{pmatrix}, \tag{4.33}$$

with spinor indices α and β, and

$$\begin{aligned}S^>_{\alpha\beta}(x,y) &= \langle \psi_{-\alpha}(x)\bar{\psi}_{+\beta}(y)\rangle, \\ S^<_{\alpha\beta}(x,y) &= -\langle \bar{\psi}_{-\beta}(y)\psi_{+\alpha}(x)\rangle, \\ S^t_{\alpha\beta}(x,y) &= \langle T\left\{\psi_{+\alpha}(x)\bar{\psi}_{+\beta}(y)\right\}\rangle \\ &= \theta(x^0 - y^0)S^>_{\alpha\beta}(x,y) + \theta(y^0 - x^0)S^<_{\alpha\beta}(x,y), \\ S^{\bar{t}}_{\alpha\beta}(x,y) &= \langle \bar{T}\left\{\psi_{-\alpha}(x)\bar{\psi}_{-\beta}(y)\right\}\rangle \\ &= \theta(x^0 - y^0)S^<_{\alpha\beta}(x,y) + \theta(y^0 - x^0)S^>_{\alpha\beta}(x,y).\end{aligned} \tag{4.34}$$

The SD equations are given by

$$\hat{S}(x,y) = \hat{S}^0(x,y) - i\int_C d^4z \int_C d^4w\, \hat{S}^0(x,z)\hat{\Sigma}(z,w)\hat{S}(w,y), \tag{4.35}$$

$$\hat{S}(x,y) = \hat{S}^0(x,y) - i\int_C d^4z \int_C d^4w\, \hat{S}(x,z)\hat{\Sigma}(z,w)\hat{S}^0(w,y), \tag{4.36}$$

where $\hat{S}^0$ represents the free propagator that follows[3]

$$(i\partial\!\!\!/_x - m)\hat{S}^0(x-y) = i\delta_C^{(4)}(x-y). \tag{4.37}$$

If $(i\partial\!\!\!/_x - m)$ and $(i\partial\!\!\!/_y + m)$ act on $\hat{S}(x, y)$

$$(i\partial\!\!\!/_x - m)\hat{S}(x,y) = i\delta_C^{(4)}(x-y)\tau^3 + \int_{-\infty}^{\infty} d^4w\ \hat{\Sigma}(x,w)\tau^3\hat{S}(w,y), \tag{4.38}$$

$$\hat{S}(x,y)(i\partial\!\!\!/_y + m) = -i\delta_C^{(4)}(x-y)\tau^3 - \int_{-\infty}^{\infty} d^4w\ \hat{S}(x,w)\tau^3\hat{\Sigma}(w,y), \tag{4.39}$$

these relations lead to

$$i\left[\partial\!\!\!/_x\hat{S}(x,y) + \hat{S}(x,y)\partial\!\!\!/_y\right] = \int_{-\infty}^{\infty} d^4w\left[\hat{\Sigma}(x,w)\tau^3\hat{S}(w,y) - \hat{S}(x,w)\tau^3\hat{\Sigma}(w,y)\right]. \tag{4.40}$$

On another front, the Noether current, which involves the transformation of $\psi \to \psi' = e^{i\alpha Q}\psi$, is

$$J^\mu(x) = -Q\bar{\psi}(x)\gamma^\mu\psi(x), \tag{4.41}$$

and its first derivative becomes

$$\begin{aligned}\partial_{x\mu}\langle J^\mu(x)\rangle &= -Q(\partial_{x\mu} + \partial_{y\mu})\left.\langle\bar{\psi}(y)\gamma^\mu\psi(x)\rangle\right|_{y\to x}\\ &= -Q\ \mathrm{Tr}\left.\langle\bar{\psi}(y)\overleftarrow{\partial\!\!\!/}_y\psi(x) + \bar{\psi}(y)\overrightarrow{\partial\!\!\!/}_x\psi(x)\rangle\right|_{y\to x}\\ &= -iQ\ \mathrm{Tr}\left.\left[i\left(\partial\!\!\!/_x + \partial\!\!\!/_y\right)S^>(x,y)\right]\right|_{y\to x}.\end{aligned} \tag{4.42}$$

[3]Free propagator for fermion satisfies

$$(i\partial\!\!\!/_x - m)\hat{S}^0(x-y) = i\delta^{(4)}(x-y),$$
$$\hat{S}^0(x-y)(i\partial\!\!\!/_y + m) = -i\delta^{(4)}(x-y),$$

where

$$i\delta^{(4)}(x-y) = \int \frac{d^4p}{(2\pi)^4}(p\!\!\!/ - m)e^{-ip\cdot(x-y)}\tilde{S}^0(p) = \int \frac{d^4p}{(2\pi)^4}(i\partial\!\!\!/_x - m)e^{-ip\cdot(x-y)}\tilde{S}^0(p),$$

$$-i\delta^{(4)}(x-y) = \int \frac{d^4p}{(2\pi)^4}(-p\!\!\!/ + m)e^{-ip\cdot(x-y)}\tilde{S}^0(p) = \int \frac{d^4p}{(2\pi)^4}e^{-ip\cdot(x-y)}\tilde{S}^0(p)(i\partial\!\!\!/_y + m).$$

Then, we finally obtain the diffusion equation for fermion

$$\frac{\partial n_\psi(X)}{\partial t_X} + \nabla_X \cdot \mathbf{j}_\psi(X) = -iQ \int_{-\infty}^{t_X} dz^0 \int_{-\infty}^{\infty} d^3\mathbf{z}\, \mathrm{Tr}\, \big[\Sigma^>(X,z)G^<(z,X) \\ -\Sigma^<(X,z)S^>(z,X) - S^>(X,z)\Sigma^<(z,X) \\ +S^<(X,Z)\Sigma^>(z,X)\big]. \qquad (4.43)$$

It should be emphasized that the right-hand side in the diffusion equation is able to be extracted by the perturbative calculations incorporating the CP-violating processes.

4.1.2 CP-Violating Source Term and Chirality-Changing Rate

In our model, relevant interactions to the BAU are

$$\mathcal{L} = \overline{\tilde{H}^0}\left(c_L^{\tilde{H}^0\tilde{S}} P_L \phi_a^0 + c_R^{\tilde{H}^0\tilde{S}} P_R \phi_b^0\right)\tilde{S} + \text{h.c.}, \qquad (4.44)$$

in any types of Z_2 assignment of new fermions. Here, a and b correspond to 1 or 2. This interaction leads to the self-energy of $\Sigma_{\tilde{H}}$ as in Fig. 4.2. Although the interaction causes a mixing between $\tilde{H}^0$ and $\tilde{S}$, the calculation for the diffusion equation is performed with gauge eigenstates. This treatment is adequate under the symmetric phase as long as the space-dependent VEVs are not so large.

The self-energy of $\Sigma_{\tilde{H}}$ is described with the free propagator for $\tilde{S}$

$$\Sigma_{\tilde{H}}(X,z) = g_{LR}(X,w)P_L\hat{S}^0_{\tilde{S}}(X_z)P_R + g_{LL}(X,z)P_L\hat{S}^0_{\tilde{S}}(X,z)P_L \\ + g_{RR}(X,z)P_R\hat{S}^0_{\tilde{S}}(X,z)P_R + g_{RL}(X,z)P_R\hat{S}^0_{\tilde{S}}(X,z)P_L, \qquad (4.45)$$

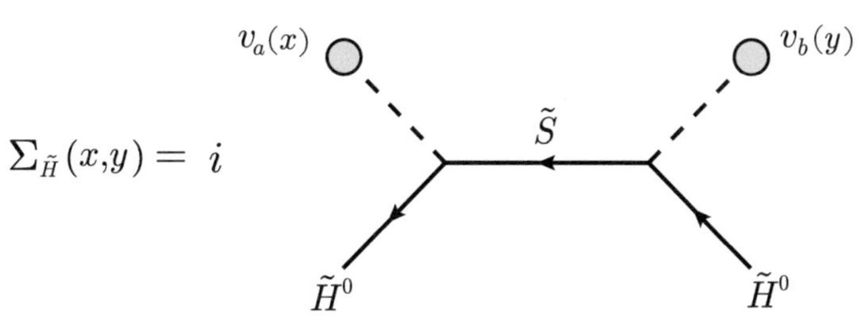

Fig. 4.2 Self-energy of $\tilde{H}^0$ that is essential for the BAU in this model

where

$$g_{LR}(X,z) = -\frac{1}{2} c_L^{\tilde{H}^0\tilde{S}} c_L^{\tilde{H}^0\tilde{S}} v_a(X) v_b(z), \tag{4.46}$$

$$g_{LL}(X,z) = -\frac{1}{2} c_L^{\tilde{H}^0\tilde{S}} \left(c_R^{\tilde{H}^0\tilde{S}}\right)^* v_a(X) v_b(z), \tag{4.47}$$

$$g_{RR}(X,z) = -\frac{1}{2} c_R^{\tilde{H}^0\tilde{S}} \left(c_L^{\tilde{H}^0\tilde{S}}\right) v_b(X) v_a(z), \tag{4.48}$$

$$g_{RL}(X,z) = -\frac{1}{2} c_R^{\tilde{H}^0\tilde{S}} \left(c_R^{\tilde{H}^0\tilde{S}}\right)^* v_b(X) v_a(z). \tag{4.49}$$

Applying the obtained $\Sigma_{\tilde{H}}$ to the diffusion equation for $\tilde{H}^0$, we obtain

$$\frac{\partial n_{\tilde{H}}(X)}{\partial t_X} + \nabla_X \cdot \mathbf{j}_{\tilde{H}}(X) = S_{\tilde{H}}(X) + \Gamma_{\tilde{H}}(X), \tag{4.50}$$

where

$$\mathcal{S}_{\tilde{H}}(X) = Q \int_{-\infty}^{t_X} dz^0 \int_{-\infty}^{\infty} d^3\mathbf{z} \sum_{A,B=L,R} i \,\mathrm{Tr}\left[\left\{g_{AB}(X,z) - g_{\alpha\beta}(z,X)\right\} \mathrm{Im}(\mathcal{G}_{AB})\right], \tag{4.51}$$

$$\Gamma_{\tilde{H}}(X) = Q \int_{-\infty}^{t_X} dz^0 \int_{-\infty}^{\infty} d^3\mathbf{z} \sum_{AB=L,R} \mathrm{Tr}\left[\left\{g_{AB}(X,z) + g_{AB}(z,X)\right\} \mathrm{Re}(\mathcal{G}_{AB})\right], \tag{4.52}$$

with

$$\mathcal{G}_{AB} = P_A S_{\tilde{S}}^{0,>}(X,z) P_B S_{\tilde{H}}^{0,<}(z,X) - P_A S_{\tilde{S}}^{0,<}(X,z) P_B S_{\tilde{H}}^{0,>}(z,X). \tag{4.53}$$

If we assume that the wall of bubbles is smooth, the derivative expansion with respect to the VEVs is valid. Then, we consider up to the first derivative, $v_{a,b}(z) \simeq v_{a,b}(X) + \partial_X^\mu v_{a,b}(X)(z-X)_\mu$, which leads to

$$g_{LL}(X,z) - g_{LL}(z,X) = \frac{\kappa_S}{2} c_L^{\tilde{H}^0\tilde{S}} \left(c_R^{\tilde{H}^0\tilde{S}}\right)^* v^2(X) \partial^\mu \beta (z-X)_\mu, \tag{4.54}$$

$$g_{RR}(X,z) - g_{RR}(z,X) = -\frac{\kappa_S}{2} c_R^{\tilde{H}^0\tilde{S}} \left(c_L^{\tilde{H}^0\tilde{S}}\right)^* v^2(X) \partial^\mu \beta (z-X)_\mu, \tag{4.55}$$

$$g_{RL}(X,z) - g_{RL}(z,X) = 0, \tag{4.56}$$

$$g_{LR}(X,z) - g_{LR}(z,X) = 0, \tag{4.57}$$

where $\kappa_S = +1$ for $(a,b) = (2,1)$, $\kappa_S = -1$ for $(a,b) = (1,2)$ and $\kappa_S = 0$ for $(a,b) = (1,1)$, $(2,2)$. As seen above, the nonzero expressions are necessarily proportional to $(z-X)_\mu$, and these features come from the difference of $v_a(X)v_b(z) - v_a(z)v_b(X)$. This situation implies that the CP-violating source term vanishes if only one doublet exists. Namely, at least two different VEVs involved in the electroweak symmetry breaking are needed for the nonzero CP-violating source term in the framework of the CTP formalism.

The final expression for the CP-violating source term is given by[4]

$$\mathcal{S}_{\tilde{H}}(X) = \kappa_S \cdot 2m_{\tilde{S}} m_{\tilde{H}} \mathrm{Im}\left[c_L^{\tilde{H}^0\tilde{S}}\left(c_R^{\tilde{H}^0\tilde{S}}\right)^*\right] v^2(X)\partial_{t_X}\beta(X)\,\mathcal{I}_{\tilde{S}\tilde{H}}, \tag{4.58}$$

where

$$\begin{aligned}\mathcal{I}_{\tilde{S}\tilde{H}} = \frac{1}{4\pi^2}\int_0^\infty \frac{k^2dk}{\omega_{\tilde{S}}\omega_{\tilde{H}}}\Big[&-2\left\{\mathrm{Im}(n_{\tilde{S}}) + \mathrm{Im}(n_{\tilde{H}})\right\} G(\omega_{\tilde{S}}, \Gamma^t_{\tilde{S}}, \omega_{\tilde{H}}, \Gamma^t_{\tilde{H}})\\ &+\left\{1 - 2\mathrm{Re}(n_{\tilde{S}})\right\} I(\omega_{\tilde{H}}, \Gamma^t_{\tilde{H}}, \omega_{\tilde{S}}, \Gamma^t_{\tilde{S}})\\ &+\left\{1 - 2\mathrm{Re}(n_{\tilde{H}})\right\} I(\omega_{\tilde{S}}, \Gamma^t_{\tilde{S}}, \omega_{\tilde{H}}, \Gamma^t_{\tilde{H}})\Big],\end{aligned} \tag{4.59}$$

with

$$I(a,b,c,d) = (b+d)\left[\frac{a+c}{\left\{(b+d)^2 + (a+b)^2\right\}^2} + \frac{a-c}{\left\{(b+d)^2 + (a-b)^2\right\}^2}\right], \tag{4.60}$$

$$G(a,b,c,d) = \frac{1}{2}\left[\frac{(a+c)^2 - (b+d)^2}{\left\{(a+c)^2 + (b+d)^2\right\}^2} - \frac{(a-c)^2 - (b+d)^2}{\left\{(a-c)^2 + (b+d)^2\right\}^2}\right]. \tag{4.61}$$

and $n_i = 1/\left(e^{(\omega_i - i\Gamma_i)/T} + 1\right)$ and $\omega_i = \sqrt{k^2 + m_i^2}$. Γ_i^t represents thermal width, whose order of the magnitude is roughly $\Gamma_i^t \sim gT$ with a typical coupling g in model. This situation results in $\mathcal{S}_{\tilde{H}} \sim \mathcal{O}(g^4)$ with an assumption of $c_L^{\tilde{H}^0\tilde{S}} \sim c_R^{\tilde{H}^0\tilde{S}} \sim g$. The CP-violating source term becomes zero if $v(X) = 0$, $\partial_{t_X}\beta = 0$, and $\mathcal{I}_{\tilde{S}\tilde{H}} = 0$, where the third condition is achieved if $\Gamma^t_{\tilde{H}} = \Gamma^t_{\tilde{S}} = 0$.

Here, we comment on the derivative expansions of the VEVs. Since only time-derivative terms are left after space integrations, the obtained expressions are proportional to v_w if we take the wall rest frame, $\bar{z} = z + v_w t$, as in Fig. 4.3. Therefore, it is appropriate that we consider up to the first order differentiation of the VEVs in the derivative expansions if $v_w < 1$. In general, the bubble velocity is conventionally required to be slower than the speed of sound, namely, $v_w \lesssim 1/\sqrt{3} \sim 0.58$. Detailed discussions are found in [6–8].

[4]We take $Q = -1$.

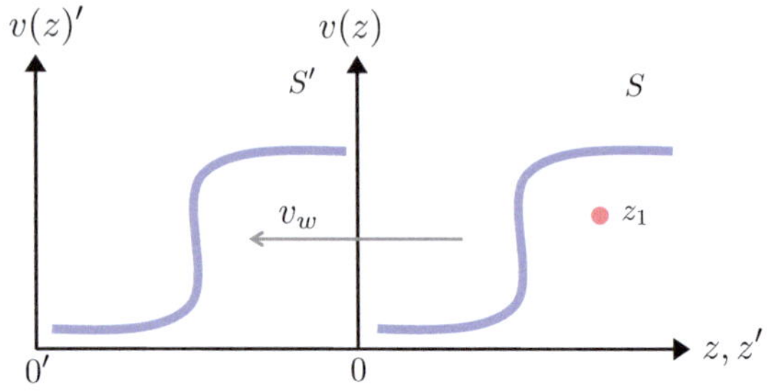

Fig. 4.3 Wall rest frame with the velocity v_w. A coordinate in S' is described by $z' = z + v_w t$

Similarly, the chirality-changing rates have the following structures:

$$g_{LL}(X, z) + g_{LL}(z, X) = c_L^{\tilde{H}^0\tilde{S}} \left(c_R^{\tilde{H}^0\tilde{S}}\right)^* v_a(X) v_b(X), \tag{4.62}$$

$$g_{RR}(X, z) + g_{RR}(z, X) = c_R^{\tilde{H}^0\tilde{S}} \left(c_L^{\tilde{H}^0\tilde{S}}\right)^* v_a(X) v_b(X), \tag{4.63}$$

$$g_{RL}(X, z) + g_{RL}(z, X) = \left|c_R^{\tilde{H}^0\tilde{S}}\right|^2 v_{a(b)}^2(X), \tag{4.64}$$

$$g_{LR}(X, z) + g_{LR}(z, X) = \left|c_L^{\tilde{H}^0\tilde{S}}\right|^2 v_{b(a)}^2(X), \tag{4.65}$$

and they lead to

$$\begin{aligned}
\Gamma_{\tilde{H}}(X) = \frac{1}{T} \Bigg[& \Bigg\{ \left(\left|c_L^{\tilde{H}^0\tilde{S}}\right|^2 v_a^2(X) + \left|c_R^{\tilde{H}^0\tilde{S}}\right|^2 v_b^2(X) \right) \mathcal{F}^+_{\tilde{S}\tilde{H}^0} \\
& + 2 v_a(X) v_b(X) m_{\tilde{H}^0} m_{\tilde{S}} \times \mathrm{Re}\left[c_L^{\tilde{H}^0\tilde{S}} \left(c_R^{\tilde{H}^0\tilde{S}}\right)^* \right] \mathcal{R}^+_{\tilde{S}\tilde{H}^0} \Bigg\} \left(\mu_{\tilde{S}} + \mu_{\tilde{H}^0}\right) \\
& + \Bigg\{ \left(\left|c_L^{\tilde{H}^0\tilde{S}}\right|^2 v_a^2(X) + \left|c_R^{\tilde{H}^0\tilde{S}}\right|^2 v_b^2(X) \right) \mathcal{F}^-_{\tilde{S}\tilde{H}^0} \\
& + 2 v_a(X) v_b(X) m_{\tilde{H}^0} m_{\tilde{S}} \times \mathrm{Re}\left[c_L^{\tilde{H}^0\tilde{S}} \left(c_R^{\tilde{H}^0\tilde{S}}\right)^* \right] \mathcal{R}^-_{\tilde{S}\tilde{H}^0} \Bigg\} \left(\mu_{\tilde{S}} - \mu_{\tilde{H}^0}\right) \Bigg] \\
\equiv {} & \Gamma^+_{\tilde{H}} \left(\mu_{\tilde{S}} + \mu_{\tilde{H}^0}\right) + \Gamma^-_{\tilde{H}} \left(\mu_{\tilde{S}} - \mu_{\tilde{H}^0}\right),
\end{aligned} \tag{4.66}$$

with

$$\mathcal{F}^\pm_{\tilde{S}\tilde{H}^0} = \frac{1}{4\pi^2} \int_0^\infty dk \, \frac{k^2}{\omega_{\tilde{S}} \omega_{\tilde{H}^0}} \left[F^\pm_{\tilde{S}\tilde{H}^0} - k^2 R^\pm_{\tilde{S}\tilde{H}^0} \right], \tag{4.67}$$

$$\mathcal{R}^\pm_{\tilde{S}\tilde{H}^0} = \frac{1}{4\pi^2} \int_0^\infty dk \, \frac{k^2}{\omega_{\tilde{S}} \omega_{\tilde{H}^0}} R^\pm_{\tilde{S}\tilde{H}^0}, \tag{4.68}$$

where

$$F^{\pm}_{\tilde{S}\tilde{H}^0}= \left\{\mathrm{Re}(\tilde{n}_{\tilde{S}}) \mp \mathrm{Re}(\tilde{n}_{\tilde{H}^0})\right\} \Big\{(\omega_{\tilde{S}}\omega_{\psi_{\tilde{H}^0}}+\Gamma^t_{\psi_{\tilde{S}}}\Gamma^t_{\tilde{H}^0})\alpha^- +(\omega_{\tilde{S}}\omega_{\psi_{\tilde{H}^0}}-\Gamma^t_{\psi_{\tilde{S}}}\Gamma^t_{\tilde{H}^0})\alpha^+$$
$$-(\omega_{\tilde{S}}\omega_{\tilde{H}^0}-\Gamma^t_{\tilde{H}^0}\Gamma^t_{\tilde{S}})\beta^- +(\omega_{\tilde{S}}\omega_{\tilde{H}^0}+\Gamma^t_{\tilde{H}^0}\Gamma^t_{\tilde{S}})\beta^+\Big\}$$
$$-\left\{\mathrm{Im}(\tilde{n}_{\tilde{S}}) \mp \mathrm{Im}(\tilde{n}_{\tilde{H}^0})\right\}\Big\{(\omega_{\tilde{S}}\omega_{\tilde{H}^0}-\Gamma^t_{\tilde{S}}\Gamma^t_{\tilde{H}^0})\beta^+ - (\omega_{\tilde{S}}\Gamma^t_{\tilde{H}^0}+\omega_{\tilde{H}^0}\Gamma^t_{\tilde{S}})\alpha^+\Big\}$$
$$-\Big\{\mathrm{Im}(n\tilde{S}) \pm \mathrm{Im}(\tilde{n}_{\tilde{H}^0})\Big\}\Big\{(\omega_{\tilde{S}}\omega_{\tilde{H}^0}+\Gamma^t_{\tilde{S}}\Gamma^t_{\tilde{H}^0})\beta^+ - (\omega_{\tilde{S}}\Gamma^t_{\tilde{H}^0}-\omega_{\tilde{H}^0}\Gamma^t_{\tilde{S}})\alpha^+\Big\},$$
$$R^{\pm}_{\tilde{S}\tilde{H}^0} = -\left\{\mathrm{Re}\left(\tilde{n}_{\tilde{S}}\right) \mp \mathrm{Re}\left(\tilde{n}_{\tilde{H}^0}\right)\right\}\left(\alpha^+_{\tilde{S}\tilde{H}^0} - \alpha^-_{\tilde{S}\tilde{H}^0}\right) + \left\{\mathrm{Im}\left(\tilde{n}_{\tilde{S}} \mp \mathrm{Im}\left(n_{\tilde{H}^0}\right)\right)\right\}\beta^+_{\tilde{S}\tilde{H}^0}$$
$$-\left\{\mathrm{Im}\left(\tilde{n}_{\tilde{S}} \pm \mathrm{Im}\left(n_{\tilde{H}^0}\right)\right)\right\}\beta^+_{\tilde{S}\tilde{H}^0}, \tag{4.69}$$

and

$$\alpha^{\pm}_{\tilde{S}\tilde{H}^0} = \frac{\Gamma^t_{\tilde{S}} + \Gamma^t_{\tilde{H}^0}}{(\omega_{\tilde{S}} \pm \omega_{\tilde{H}^0})^2 + (\Gamma^t_{\tilde{S}} + \Gamma^t_{\tilde{H}^0})^2}, \quad \beta^{\pm}_{\tilde{S}\tilde{H}^0} = -\frac{\omega_{\tilde{S}} \pm \omega_{\tilde{H}^0}}{(\omega_{\tilde{S}} \pm \omega_{\tilde{H}^0})^2 + (\Gamma^t_{\tilde{S}} + \Gamma^t_{\tilde{H}^0})^2}. \tag{4.70}$$

Note that there is no vector current for $\tilde{S}$ since it is a Majorana fermion, and it follows that the corresponding chemical potential also vanishes. This situation makes the chirality-changing rate a bit simple:

$$\Gamma_{\tilde{H}}(X) = -\left(\Gamma^-_{\tilde{H}} - \Gamma^+_{\tilde{H}}\right)\mu_{\tilde{H}^0} \tag{4.71}$$

These expressions are consistent with those in Ref. [9]. It is possible to fix phases of $c_L^{\tilde{H}^0\tilde{S}}$ and $c_R^{\tilde{H}^0\tilde{S}*}$ so that $\Gamma_{\tilde{H}}$ becomes minimum, and it yields more baryon numbers.

Let us define k factors

$$k_{b,f}\left(\frac{m}{T}\right) = \frac{6n_{b,f}}{T^2\mu} = \frac{6g}{T^2\mu}\int\frac{d^3k}{(2\pi)^3}\left[\frac{1}{e^{(\omega-\mu)/T}\mp 1} - \frac{1}{e^{(\omega+\mu)/T}\mp 1}\right]. \tag{4.72}$$

With the k factors, the chirality-changing rate is given by

$$\Gamma_{\tilde{H}}(X) = -\left(\Gamma^-_{\tilde{H}} - \Gamma^+_{\tilde{H}}\right)\frac{6}{T^2}\frac{n_{\tilde{H}^0}}{k_{\tilde{H}^0}} \equiv -\bar{\Gamma}_{\tilde{H}}(X)\frac{n_{\tilde{H}}}{k_{\tilde{H}^0}}, \tag{4.73}$$

with

$$\bar{\Gamma}_{\tilde{H}}(X) = \frac{6}{T^2}\left(\Gamma^-_{\tilde{H}} - \Gamma^+_{\tilde{H}}\right). \tag{4.74}$$

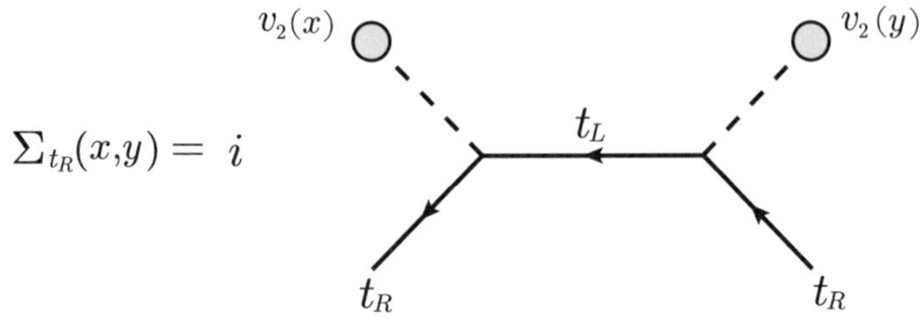

Fig. 4.4 The self-energy for the right-handed top quark that comes from the top mass interaction

4.1.3 Chirality-Changing Rate for the Top Mass Interaction

The diffusion equation for the top quark is given by

$$\frac{\partial n_t}{\partial t_X} + \nabla_X \cdot \mathbf{j}_t(X) = \Gamma_{m_t}(X) + \Gamma_Y(X). \tag{4.75}$$

Here, while $\Gamma_{m_t}(X)$ comes from the top mass interaction as in Fig. 4.4, $\Gamma_Y(X)$ corresponds to the chirality-changing rate for the top Yukawa interaction. The derivations for $\Gamma_{m_t}(X)$ and $\Gamma_Y(X)$ are much complicated compared to the calculations in the previous section. For chiral fermions, their masses coming from the VEV become zero in the symmetric phase, and we cannot ignore thermal effects on its propagator. As discussed in [10], there are new four poles, which correspond to collective fermionic modes, in the propagator of the chiral fermion. For more details, see Refs. [11–16].

Let us see the self-energy for the right-handed top quark that results from the top mass interaction in Fig. 4.4. After a bit intricate calculations, the chirality-changing rate is obtained by

$$\Gamma_{m_{t_{R(L)}}} = \Gamma^{+}_{t_{R(L)}}(X)(\mu_{t_{L(R)}} + \mu_{t_{R(L)}}) + \Gamma^{-}_{t_{R(L)}}(X)(\mu_{t_{L(R)}} - \mu_{t_{R(L)}}), \tag{4.76}$$

with

$$\Gamma^{\pm}_{t_{R(L)}} = \frac{y_t^2 v_2^2(X)}{2\pi^2 T} \int_0^\infty dk\, k^2 \mathrm{Im}\left[\left(\tilde{n}^p_{L(R)} \mp \tilde{n}^p_{R(L)}\right) \frac{Z^p_L Z^p_R}{\mathcal{E}^p_L + \mathcal{E}^p_R} + \left(\tilde{n}^p_{L(R)} \mp \tilde{n}^{h*}_{R(L)}\right) \frac{Z^p_{L(R)} Z^{h*}_{R(L)}}{\mathcal{E}^p_{L(R)} - \mathcal{E}^{h*}_{R(L)}}\right] \tag{4.77}$$

where $\mathcal{E}^p(k) = E^p(k) - i\Gamma^p(T)$ and $\mathcal{E}^h = E^h(k) - i\Gamma^h(T)$ are complex energies for particle and hole excitations, respectively. $\Gamma^{p,h}(T)$ is a thermal width for them. $Z^{p,h}$ are residues

$$Z^{p,h}_{L,R}(k) = \frac{\left(E^{p,h}(k)\right)^2 - k^2}{2m_t^2}. \tag{4.78}$$

In the limit of $t_L - t_R$ symmetry, it is obtained that $\Gamma^+_{t_R} = \Gamma^+_{t_L} = 0$ and $\Gamma^-_{t_R} = \Gamma^-_{t_L}$. This leads to the simple expressions

$$\Gamma_{m_{t_R}} = \Gamma_{t_R}(X) = \Gamma^-_{t_R}(\mu_{t_L} - \mu_{t_R}) = \Gamma^-_{t_R}\frac{6}{T^2}\left(\frac{n_{t_L}}{k_{t_L}} - \frac{n_{t_R}}{k_{t_R}}\right) \equiv \Gamma_M\left(\frac{n_{t_L}}{k_{t_L}} - \frac{n_{t_R}}{k_{t_R}}\right), \tag{4.79}$$

$$\Gamma_{m_{t_L}} = \Gamma_{t_L}(X) = \Gamma^-_{t_R}(\mu_{t_R} - \mu_{t_L}) = -\Gamma^-_{t_R}\frac{6}{T^2}\left(\frac{n_{t_L}}{k_{t_L}} - \frac{n_{t_R}}{k_{t_R}}\right) \equiv -\Gamma_M\left(\frac{n_{t_L}}{k_{t_L}} - \frac{n_{t_R}}{k_{t_R}}\right), \tag{4.80}$$

with $\Gamma_M = 6\Gamma^-_{t_R}/T^2$.

4.2 Estimation of the Baryon Number

We include the contributions from the left- and right-handed top quarks and the Higgs sector in the diffusion equations. Their number densities are defined as[5]

$$Q(X) = n_{t_L}, \tag{4.81}$$

$$T(X) = n_{t_R}, \tag{4.82}$$

$$H(X) = n_{\Phi_1} + n_{\Phi_2} + n_{\tilde{H}}, \tag{4.83}$$

and they obey the following diffusion equations:

$$\frac{\partial Q}{\partial t_X} + \nabla_X \cdot \mathbf{j}_Q(X) = \Gamma_M\left(\frac{T}{k_T} - \frac{Q}{k_Q}\right) + \Gamma_Y\left(\frac{T}{k_T} - \frac{H}{k_H} - \frac{Q}{k_Q}\right) - 2\Gamma_{ss}\left(\frac{2Q}{k_Q} - \frac{T}{k_T} - \frac{9(Q+T)}{k_B}\right), \tag{4.84}$$

$$\frac{\partial T}{\partial t_X} + \nabla_X \cdot \mathbf{j}_T(X) = -\Gamma_M\left(\frac{T}{k_T} - \frac{Q}{k_Q}\right) - \Gamma_Y\left(\frac{T}{k_T} - \frac{H}{k_H} - \frac{Q}{k_Q}\right) + 2\Gamma_{ss}\left(\frac{2Q}{k_Q} - \frac{T}{k_T} - \frac{9(Q+T)}{k_B}\right), \tag{4.85}$$

$$\frac{\partial H}{\partial t_X} + \nabla_X \cdot \mathbf{j}_H(X) = -\Gamma_Y\left(\frac{Q}{k_Q} + \frac{H}{k_H} - \frac{T}{k_T}\right) - \bar{\Gamma}_{\tilde{H}}\frac{H}{k_{\tilde{H}^0}} + S_{\tilde{H}}. \tag{4.86}$$

[5]There is no number density of the singlet in $H(X)$ since it vanishes due to the real field.

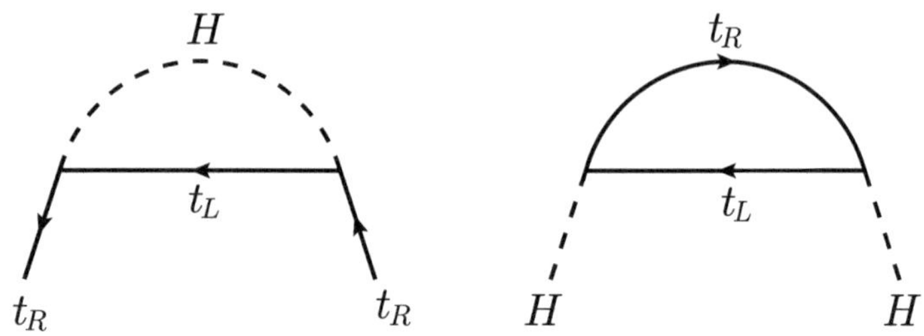

Fig. 4.5 Yukawa interactions taken in the diffusion equations

Here, we also include both the top Yukawa interaction Γ_Y in Fig. 4.5 and strong sphaleron process in the diffusion equations. The strong sphaleron results from the triangle anomaly of the axial vector current $\bar{q}\gamma^\mu\gamma_5 q$, and the fact of $\pi_3(SU(N_C)) = Z$ for $N_C \geq 2$ implies a topological transitions as well as the $SU(2)$ sphaleron. The strong sphaleron process catalyzes the number densities of the left- and right-handed quarks, and the process is expressed by a relation of the chemical potentials $\sum_{i=1}^{3}\left(\mu_{u_L^i} - \mu_{u_R^i} + \mu_{d_L^i} - \mu_{d_R^i}\right)$. For more details, see Ref. [1].

Based on estimations in Ref. [17], we assume that $\Gamma_Y,\ \Gamma_{ss} \gg \Gamma_M$. This situation implies that Γ_Y and Γ_{ss} approach to the equilibrium, which results in the same chemical potentials before and after interactions of the top Yukawa interaction and the strong sphaleron, namely, $\mu_T = \mu_H + \mu_Q$ and $\sum_{i=1}^{3}\left(\mu_{u_L^i} - \mu_{u_R^i} + \mu_{d_L^i} - \mu_{d_R^i}\right) = 0$. If the time scale of the interaction is faster than that of the bubble speed, it is regarded as equilibrium. Thus, defining δ_Y and δ_{ss} as

$$\frac{T}{k_T} - \frac{H}{k_H} - \frac{Q}{k_Q} \sim \mathcal{O}\left(\frac{1}{\Gamma_Y}\right) \equiv \delta_Y, \tag{4.87}$$

$$\frac{2Q}{k_Q} - \frac{T}{k_T} - \frac{Q}{k_Q} \sim \mathcal{O}\left(\frac{1}{\Gamma_{ss}}\right) \equiv \delta_{ss}, \tag{4.88}$$

we obtain

$$\frac{T}{k_T} - \frac{Q}{k_Q} = \frac{H}{k_H}, \tag{4.89}$$

$$2T + Q = \frac{4k_B k_T + 9k_Q k_T + k_Q k_B}{k_B + 9k_Q + 9k_T}\frac{H}{k_H}. \tag{4.90}$$

These relations are utilized for obtaining one combined diffusion equation

$$\frac{\partial H}{\partial t_X} + \frac{a}{a+b}\nabla_X\left(2\mathbf{j}_X + \mathbf{j}_Q + \mathbf{j}_H\right) + \frac{a}{a+b}\frac{\Gamma_M + \bar{\Gamma}_{\tilde{H}}}{k_H}H - \frac{a}{a+b}S_{\tilde{H}} + \mathcal{O}\left(\delta_Y, \delta_{ss}\right) = 0, \tag{4.91}$$

with $a = k_H(9k_Q + 9k_T + k_B)$ and $b = 9k_Qk_T + k_Qk_B + 4k_Tk_B$. It implies that Eqs. (4.84), (4.85), and (4.86) are converted into the single diffusion equation for H. When diffusion approximations known as the Fick diffusion law are introduced

$$\mathbf{j}_T = -D_T\nabla_X T, \qquad \mathbf{j}_Q = -D_Q\nabla_X Q, \qquad \mathbf{j}_H = -D_H\nabla_X H, \tag{4.92}$$

where D_T, D_Q, and D_H are the diffusion constants of each particle, more simple expression is obtained:

$$\frac{\partial H}{\partial t_X} - \bar{D}\nabla_X^2 H - \bar{S} + \mathcal{O}(\delta_Y, \delta_{\rm ss}) = 0, \tag{4.93}$$

with

$$\bar{D} = \frac{bD_Q + aD_H}{a+b}, \tag{4.94}$$

$$\bar{\Gamma} = \frac{1}{a+b}\left[\frac{a}{k_H}\left(\Gamma_M + \bar{\Gamma}_{\tilde{H}}\right)\right], \tag{4.95}$$

$$\bar{S} = \frac{a}{a+b}S_{\tilde{H}}. \tag{4.96}$$

Here, as in Ref. [18], we take $D_Q = D_T$ for simplicity. Since the wall thickness is much smaller than its radius, it would be valid to focus on only one direction that the wall moves. Then, we rewrite the above equation in the wall rest frame of $\bar{z} = z + v_w t$

$$v_w H'(\bar{z}) - \bar{D}H''(\bar{z}) + \bar{\Gamma}H(\bar{z}) - \bar{S}(\bar{z}) + \mathcal{O}(\delta_Y, \delta_{\rm ss}) = 0. \tag{4.97}$$

While $\bar{z} < 0$ corresponds to the symmetric phase, $\bar{z} > 0$ is in the broken phase. Assuming that $\bar{\Gamma}$ is nonzero and constant for $\bar{z} > 0$, one obtain a solution for H in the symmetric phase

$$H(\bar{z}) = \mathcal{A}e^{v_w\bar{z}/\bar{D}} \tag{4.98}$$

with

$$\mathcal{A} = \frac{1}{\bar{D}\lambda_+}\int_0^\infty dz'\bar{S}(z')e^{-\lambda_+ z'}, \qquad \lambda_+ = \frac{v_w + \sqrt{v_w^2 + 4\bar{D}\bar{\Gamma}}}{2\bar{D}}. \tag{4.99}$$

In the case in which we neglect contribution from leptons, the left-handed number density of n_L becomes the sum of the left-handed quark densities. Since the first- and second-generation quarks are only generated by the strong sphaleron, it is possible to relate the left-handed number densities of the first- and second-generation quarks, Q_1 and Q_2, to those of the third-generation quarks Q and T. Based on Refs. [17, 19], it is obtained that $n_L = Q_1 + Q_2 + Q = 5Q + 4T$, and the left-handed number density is finally expressed by

$$n_L = 5Q(\bar{z}) + 4T(\bar{z}) = -\left[r_1 + \frac{r_2 v_w^2}{\Gamma_{ss}\bar{D}}\left(1 - \frac{D_Q}{\bar{D}}\right)\right] H(\bar{z}), \tag{4.100}$$

where

$$r_1 = \frac{9k_Q k_T - 5k_Q k_B - 8k_T k_B}{a}, \tag{4.101}$$

$$r_2 = \frac{k_H k_B^2 (5k_Q + 4k_T)(k_Q + 2k_T)}{a^2}. \tag{4.102}$$

It is seen that n_L is proportional to H, which implies the proportion to the CP-violating source term $\bar{S}$ as well. With these values, the diffusion equation for the baryon number can be obtained

$$D_Q n_B''(\bar{z}) - v_w n_B'(\bar{z}) - \theta(-\bar{z})\mathcal{R} n_B(\bar{z}) = \theta(-\bar{z})\frac{N_g}{2}\Gamma_B^{(s)} n_L(\bar{z}), \tag{4.103}$$

where $\mathcal{R} = (15/4)\Gamma_B^{(s)}$ that is called a relaxation term. Finally, the expression for n_B [3, 4, 9, 10, 17] can be found to be

$$n_B(\bar{z} > 0) = \frac{-N_g \Gamma_B^{(s)}}{2D_Q \Lambda_+} \int_{-\infty}^{0} dz' n_L(z') e^{-\Lambda_- z'} \tag{4.104}$$

with

$$\Lambda_\pm = \frac{v_w \pm \sqrt{v_w^2 + 4\mathcal{R} D_Q}}{2D_Q}. \tag{4.105}$$

What is important is that the above expression of n_B includes Sakharov's criteria, i.e.,

1. Baryon number violating process: $\Gamma_B^{(s)}$,
2. CP violation: n_L,
3. Out of equilibrium: v_w.

It should be emphasized that the final baryon number can not be produced if one of them is missing. Moreover, it depends on the sphaleron rate in the broken phase whether the baryon number is finally left in the Universe.

References

1. A. Riotto, hep-ph/9807454
2. K.C. Chou, Z.B. Su, B.L. Hao, L. Yu, Phys. Rept. **118**, 1 (1985)
3. A. Riotto, Nucl. Phys. B **518**, 339 (1998). [hep-ph/9712221]
4. A. Riotto, Phys. Rev. D **58**, 095009 (1998). [hep-ph/9803357]
5. S. Iso, K. Shimada, M. Yamanaka, JHEP **1404**, 062 (2014). [arXiv: 1312.7680 [hep-ph]]
6. J.M. No, Phys. Rev. D **84**, 124025 (2011). [arXiv: 1103.2159 [hep-ph]]
7. C. Caprini, J.M. No, JCAP **1201**, 031 (2012). [arXiv: 1111.1726 [hep-ph]]
8. J. Kozaczuk, JHEP **1510**, 135 (2015). [arXiv:1506.04741 [hep-ph]]
9. E. Senaha, Phys. Rev. D **88**(5), 055014 (2013). [arXiv: 1308.3389 [hep-ph]]
10. C. Lee, V. Cirigliano, M.J. Ramsey-Musolf, Phys. Rev. D **71**, 075010 (2005). [hep-ph/0412354]
11. H.A. Weldon, Phys. Rev. D **26**, 2789 (1982)
12. H.A. Weldon, Phys. Rev. D **40**, 2410 (1989)
13. V.V. Klimov, Sov. J. Nucl. Phys. **33**, 934 (1981), [Yad. Fiz. **33**, 1734 (1981)]
14. V.V. Klimov, Sov. Phys. JETP **55**, 199 (1982), [Zh. Eksp. Teor. Fiz. **82**, 336 (1982)]
15. H.A. Weldon, Phys. Rev. D **61**, 036003 (2000). [hep-ph/9908204]
16. M. Le Bellac, *Thermal Field Theory* (Cambridge University Press, Cambridge, 1996)
17. P. Huet, A.E. Nelson, Phys. Rev. D **53**, 4578 (1996). [hep-ph/9506477]
18. D.J.H. Chung, B. Garbrecht, M.J. Ramsey-Musolf, S. Tulin, JHEP **0912**, 067 (2009). [arXiv: 0908.2187 [hep-ph]]
19. A.G. Cohen, D.B. Kaplan, A.E. Nelson, Phys. Lett. B **336**, 41 (1994). [hep-ph/9406345]

Chapter 5
Electric Dipole Moments

Abstract Electric dipole moments are quantities which break time-reversal and parity symmetry. It implies CP violation assuming that CPT theorem. The search for the electric dipole moments is strong tools for examinations of electroweak baryogenesis. In this chapter, we discuss the relationship between the quantities and the CP-violating process for the BAU and see how they work in the examination.

Keywords Electric dipole moments · Barr-Zee diagrams · CP-violating source term

5.1 Electric Dipole Moments

The electric dipole moment is a quantity defined as

$$\mathbf{d} = d\frac{\mathbf{s}}{|\mathbf{s}|} \tag{5.1}$$

where $\mathbf{s}$ is spin, and this quantity is sometimes explained as electric bias inside a particle. An interaction with electric field $\mathbf{E}$ is described by

$$\mathcal{H}_{\mathrm{EDM}} = -\mathbf{d}\cdot\mathbf{E}. \tag{5.2}$$

Considering parity (P) and time-reversal (T) transformations, we can see that the Hamiltonian changes a sign

$$\mathbf{d}\cdot\mathbf{E} \xrightarrow{\mathrm{P}} -\mathbf{d}\cdot\mathbf{E}, \qquad \mathbf{d}\cdot\mathbf{E} \xrightarrow{\mathrm{T}} -\mathbf{d}\cdot\mathbf{E}, \tag{5.3}$$

This chapter is based on the following article: Kaori Fuyuto, Junji Hisano, and Eibun Senaha, Phys. Lett. B **755**, 491 (2016), Copyright ©2016 The Authors. Published by Elsevier B.V.

© Springer Nature Singapore Pte Ltd. 2018
K. Fuyuto, *Electroweak Baryogenesis and Its Phenomenology*, Springer Theses,
https://doi.org/10.1007/978-981-13-1008-9_5

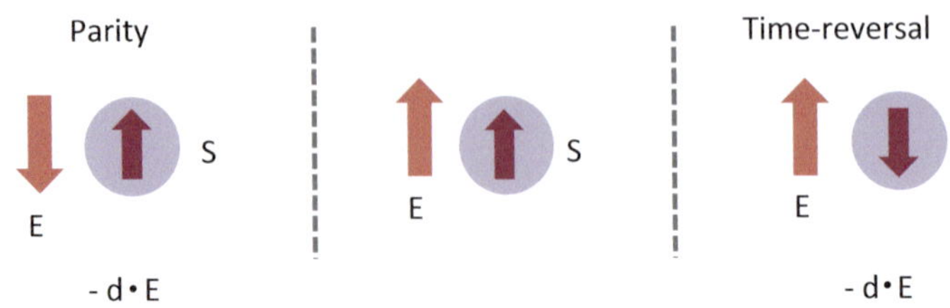

Fig. 5.1 Features of the EDMs under parity and time-reversal transformations. An arrow inside a particle indicates the spin, and the other arrow corresponds to the electric field in middle figure. Left figure shows that the parity transformation changes the arrow of the electric field changes. Right figure describes that the direction of the spin becomes opposite under the time-reversal transformation

because the electric field (spin) changes the sign under P (T) transformations as in Fig. 5.1. If we assume the CPT invariance, T violation is equivalent to that of CP symmetry. Therefore, the EDM is known as the CP-violating quantity. For reviews on the EDMs, see [1–5].

The Lagrangian for a relativistic particle is given by

$$\mathcal{L}_{\text{EDM}} = -d\frac{i}{2}\bar{\psi} F_{\mu\nu}\sigma^{\mu\nu}\gamma_5\psi. \tag{5.4}$$

From the interaction, it is seen that the EDM is originated from the CP-violating flavor-diagonal process. The CP phase in the CKM matrix is accompanied with the flavor-violating process; therefore it is not so simple to induce the EDM. Previous studies [6–10] have shown that the quark EDMs induced by the KM phase appear at three-loop level, and its order is roughly $d_d \sim 10^{-34}\ e \cdot \text{cm}$ for the down quark. Moreover, that of the electron is $d_e \sim 10^{-40}\ e \cdot \text{cm}$ [11]. These values are undetectably small; in other words, they are the background-free quantities in terms of looking for new physics that may induce the sizable EDMs. Equation (5.4) also implies that the EDMs are chirality-flipping processes. Considering naive dimensional analysis, we find that

$$d_e \sim e\frac{m_e}{\Lambda^2} = C \times 10^{-23} \times \left(\frac{1\text{TeV}}{\Lambda}\right)^2, \tag{5.5}$$

$$d_d \sim e\frac{m_d}{\Lambda^2} = C \times 10^{-22} \times \left(\frac{1\text{TeV}}{\Lambda}\right)^2, \tag{5.6}$$

where C represents a loop factor $\sim 10^{-(2-4)}$.

Table. 5.1 shows current limits and future sensitivities of the EDMs of various species [19].[1] The origins of the EDMs are different from each other; therefore the measurements of those of the various species are important for checking new CP-

[1]The limit on the electron EDM is obtained by assuming that semileptonic four-fermion interactions do not contribute to system. For more details, see the recent works [17, 18, 20–22].

Table 5.1 Limits on the EDMs of various species. The future sensitivity of the mercury remains the same

Particle species	Current limit	Expected limit
Neutron (n)	$\lvert d_n \rvert < 3.0 \times 10^{-26}\ e \cdot \mathrm{cm}$ (90% C.L.) [12]	$1.0 \times 10^{-28}\ e \cdot \mathrm{cm}$ [5, 13]
Proton (p)	–	$1.0 \times 10^{-29}\ e \cdot \mathrm{cm}$ [13]
Mercury (Hg)	$\lvert d_{\mathrm{Hg}} \rvert < 7.4 \times 10^{-30}\ e \cdot \mathrm{cm}$ (95% C.L.) [14]	$7.4 \times 10^{-30}\ e \cdot \mathrm{cm}$
Radium (Ra)	$\lvert d_{\mathrm{Ra}} \rvert < 5.0 \times 10^{-22}\ e \cdot \mathrm{cm}$ (95% C.L.) [15]	$1.0 \times 10^{-30}\ e \cdot \mathrm{cm}$ [13]
Electron (e)	$\lvert d_e \rvert < 8.7 \times 10^{-29}\ e \cdot \mathrm{cm}$ (90% C.L.) [16, 17]	$5.0 \times 10^{-30}\ e \cdot \mathrm{cm}$ [13, 18]

violating source. In addition, it can occur that the EDM of one particle completely disappears due to a cancellation of the CP phase if new physics include more than one phase. In such a case, the other EDMs would play a more essential role in probing scenarios.

In terms of the examination of new CP violation, the measurement of the dipole operator of the top quark at collider experiments has also been studied [23–28], because the top quark might be most sensitive to new physics due to the sizable Yukawa coupling. In addition, considering that the EDMs of the light fermions are able to be induced by that of the top quark through the electroweak running [19, 29–32], we find that $|d_t| \lesssim 10^{-19}$ e cm with the new physics scale $\Lambda = 1$ TeV [32].

5.2 Two-Loop Barr-Zee Diagram

In our model, the gauge interactions for the new fermions are given by

$$\mathcal{L} = \frac{g_2}{\sqrt{2}} \left(\overline{\tilde{H}^+} \gamma^\mu \tilde{H}^0 W_\mu^+ + \overline{\tilde{H}^0} \gamma^\mu \tilde{H}^+ W_\mu^- \right) - e \overline{\tilde{H}^+} \gamma^\mu \tilde{H}^+ A_\mu, \tag{5.7}$$

which can induce the light fermion EDMs via the two-loop Barr-Zee diagram [33–38] as in Fig. 5.2. What is important is that the diagram in Fig. 4.2 partly exists in the Barr-Zee diagram, namely, the CP-violating source term is directly connected to the fermion EDMs. Employing a mass-insertion method for a comparison with the CP-violating source term,[2] the fermion EDMs are expressed by

$$\begin{aligned} \frac{d_f^{WW}}{e} &= \mp \frac{\alpha_{\mathrm{em}}^2}{64\pi^2 s_W^4} \frac{m_f m_{\tilde{H}^\pm} m_{\tilde{S}} v_a v_b}{m_W^4} \mathrm{Im} \left[c_L^{\tilde{H}^0 \tilde{S}} \left(c_R^{\tilde{H}^0 \tilde{S}} \right)^* \right] F_{WW} \left(r_{\tilde{H}^0}, r_{\tilde{S}} \right) \\ &\equiv C_{\mathrm{EDM}}^{WW} \mathrm{Im} \left[c_L^{\tilde{H}^0 \tilde{S}} \left(c_R^{\tilde{H}^0 \tilde{S}} \right)^* \right], \end{aligned} \tag{5.8}$$

[2] We checked the validity of using the mass-insertion method, and the difference between the mass-insertion method and the full calculation is less than 5%.

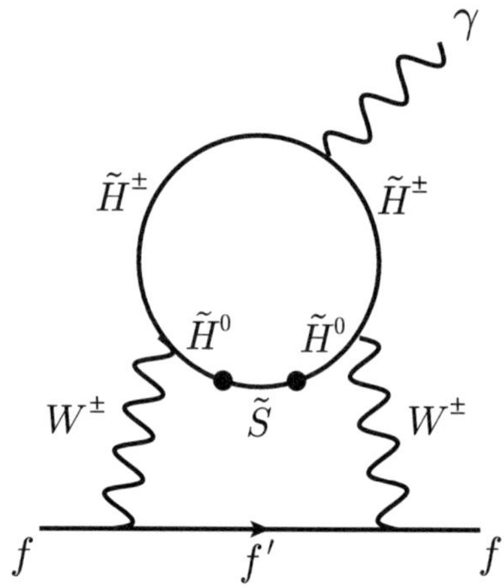

Fig. 5.2 The Barr-Zee diagram with the BAU-related CP phases

where $\alpha_{\rm em}^2 = e^2/4\pi$ and m_i is a mass of particle i. The function F_{WW} is defined as

$$F_{WW}(r_i, r_j) = \frac{f_{WW}(r_i, r_+) - f_{WW}(r_j, r_+)}{m_i^2 - m_{\tilde{H}^{\pm}}^2}, \tag{5.9}$$

with

$$f_{WW}(r_i, r_+) = \int_0^1 \frac{dx}{1-x}\frac{1}{X-1}\log X, \qquad X = \frac{(1-x)r_i + xr_+}{x(1-x)}, \tag{5.10}$$

with $r_i = m_i^2/m_W^2$, $r_j = m_j^2/m_W^2$ and $r_+ = m_{\tilde{H}^{\pm}}^2/m_W^2$. In the expressions, the negative (positive) sign corresponds to the case for up-type (down-type) fermion.

In order to understand a behavior of d_f^{WW}, let us consider the electron EDM. In Fig. 5.3, we plot the ratio $|d_e^{WW}|/d_e^{\rm exp}(= 8.7\times 10^{-29}\ e\cdot{\rm cm})$ as a function of $m_{\tilde{H}^0}$ and $m_{\tilde{S}}$. In the plot, the fixed mass is set to 300 GeV. It is seen that, while the pink line behaves as $m_e m_{\tilde{S}}/m_{\tilde{H}^0}^3$ in the heavy limit of $m_{\tilde{H}^0}$ with $m_{\tilde{S}} = 300$ GeV, the blue line is described by $m_e/(m_{\tilde{S}} m_{\tilde{H}^0})$ for the large mass limit of $\tilde{S}$ with $m_{\tilde{H}^0} = 300$ GeV.

5.3 Relationship Between the CP-Violating Source Term and EDMs

Parameterizing the CP-violating source term as

$$S_{\tilde{H}}(X) = C_{\rm BAU}\,{\rm Im}\left[c_L^{\tilde{H}^0\tilde{S}}\left(c_R^{\tilde{H}^0\tilde{S}}\right)^*\right], \tag{5.11}$$

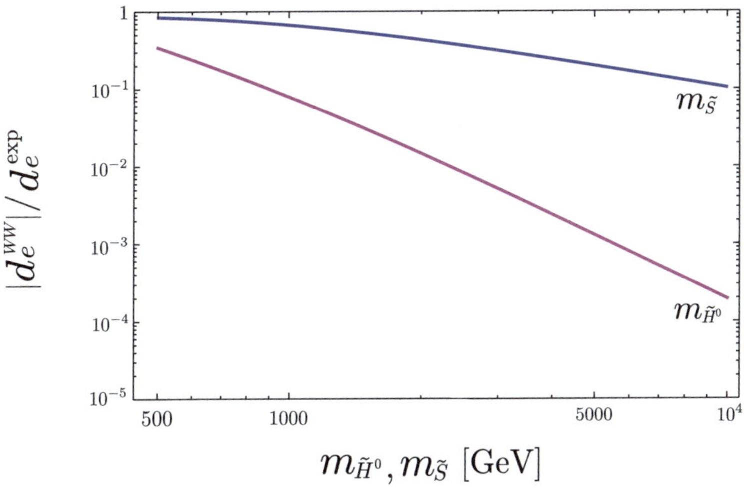

Fig. 5.3 $|d_e^{WW}|/d_e^{\rm exp}$ as a function of $m_{\tilde{H}^0}$ and $m_{\tilde{S}}$

we can replace the imaginary part with $C_{\rm EDM}^{WW}$

$$S_{\tilde{H}}(X) = \frac{C_{\rm BAU}}{C_{\rm EDM}^{WW}} \left(\frac{d_f^{WW}}{e} \right). \tag{5.12}$$

Since $C_{\rm BAU}$ contains the rather model-dependent parameters of $v^2(X)\partial_{t_X}\beta(X)$, we get rid of it and define the following parameter:

$$\bar{S} = \frac{C_{\rm BAU}}{v^2(X)\partial_{t_X}\beta(X) C_{\rm EDM}^{WW}} \left(\frac{d_e^{WW}}{e} \right)_{\rm exp}, \tag{5.13}$$

where we use the experimental limit of the electron EDM.

In Fig. 5.4, the function $\bar{S}$ is plotted as a function of $m_{\tilde{H}^0}$ and $m_{\tilde{S}}$, focusing on Type-B for Z_2 assignment for new fermions. We take $\tan\beta = 1$, and the fixed mass is set to 500 GeV. Here, the orange line corresponds to $\bar{S}$ against $m_{\tilde{H}^0}$, and the red line is that against $m_{\tilde{S}}$. One peak is seen around $m_{\tilde{H}^0} \sim m_{\tilde{S}}$, which results from the CP-violating source term. This situation is understood as follows: if a bubble wall carries low momentum and the injected momentum is approximated by $m_{\tilde{H}^0}^2$, the enhancement in the propagator of $\tilde{S}$ is caused at $m_{\tilde{S}} \sim m_{\tilde{H}^0}$. One more significant behavior of $\bar{S}$ can be read in the decoupling limit of $m_{\tilde{S}}$ and $m_{\tilde{H}^0}$. While $\bar{S}$ grows in the large mass limit of $m_{\tilde{H}^0}$, it becomes flat in the heavy mass region of $m_{\tilde{S}}$. In Fig. 5.3, it is seen that the electron EDM drastically dumps in the large mass limit of $m_{\tilde{H}}$; therefore, $\bar{S}$ becomes large due to the extremely small $C_{\rm EDM}^{WW}$ in the decoupling limit. Note that, since d_e^{WW} is fixed in this estimation, the small

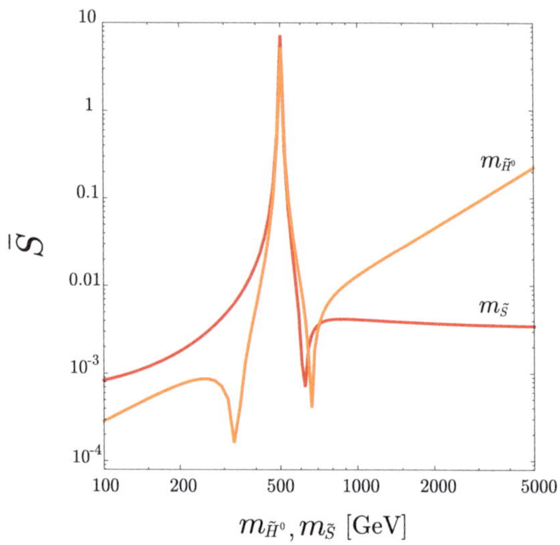

Fig. 5.4 $\bar{S}$ as a function of $m_{\tilde{H}^0}$ and $m_{\tilde{S}}$. We set $\tan\beta = 1$, and the fixed mass is 500 GeV. This is studied in a paper [39]. (Copyright ©2016 The Authors. Published by Elsevier B.V.)

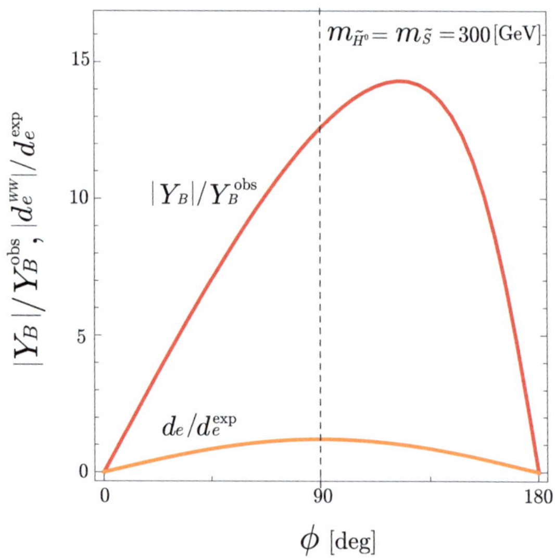

Fig. 5.5 Dependence on the BAU-related CP phase in the BAU and the electron EDM. We set $m_{\tilde{H}^0} = m_{\tilde{S}} = 300$ GeV, and $\tan\beta = 1$

value of $C_{\rm EDM}^{WW}$ corresponds to large $\mathrm{Im}\left[c_L^{\tilde{H}^0\tilde{S}}\left(c_R^{\tilde{H}^0\tilde{S}}\right)^*\right]$. Numerically, it is found that $\mathrm{Im}\left[c_L^{\tilde{H}^0\tilde{S}}\left(c_R^{\tilde{H}^0\tilde{S}}\right)^*\right] > 1$ for $m_{\tilde{H}^0} \gtrsim 1$ TeV.

As seen in the previous chapter, the baryon number is roughly expressed by

$$n_B \sim \kappa_B \frac{S_{\tilde{H}}}{\sqrt{\Gamma_t + \Gamma_{\tilde{H}}}}. \tag{5.14}$$

The final baryon number is linearly proportional to the CP-violating source term; however, the relationship between n_B and d_f^{WW} is a bit tricky since the chirality-changing rates also depend on CP phase. Figure 5.5 shows dependence on the CP phase in the BAU and the EDM of electron. Here, it is defined that $Y_B \equiv n_B/s$ and $\phi = -(\phi_{\tilde{H}} + \phi_{\tilde{S}})$. We also take $m_{\tilde{H}^0} = m_{\tilde{S}} = 300\,\text{GeV}$ and $\tan\beta = 1$. While the red line corresponds to Y_B/Y_B^{obs} with the observables in [40], the orange line is the value of $d_e^{WW}/d_e^{\text{exp}}$. It is clear that there are different dependences on the CP phase between the BAU and the electron EDM. The EDMs are absolutely proportional to $\sin\phi$ as seen in their formulas; however, the BAU does not. Also, the dependence on CP phase in the BAU varies as the masses of new fermions change.

5.4 Other Barr-Zee Diagrams

Besides the diagram in Fig. 5.2, our model can also induce another three Barr-Zee diagrams.[3] One of them is mediated by the charged scalars and W boson as in Fig. 5.6. The relevant interactions are those between the charged scalars and new fermions in the second terms of Eqs. (3.12) and (3.13). They are related to the first terms in Eqs. (3.12) and (3.13) due to the $SU(2)$ symmetry; therefore the diagram is induced by the BAU-related CP phase. However, it is expected that the heavy scalar contribution leads to a sub-leading effect on the fermion EDMs compared to Fig. 5.2.

The rest of diagrams are originated from the BAU-unrelated CP phase. As seen in Eqs. (3.12) and (3.13), the type A and B contain interactions with a real singlet h_S

$$\mathcal{L}_{\text{fer}}^{\text{A,B}} \ni -\overline{\tilde{H}^+}\left(g^S_{\tilde{H}\tilde{H}S} + i\gamma_5 g^P_{\tilde{H}\tilde{H}S}\right)\tilde{H}^+ h_S. \tag{5.15}$$

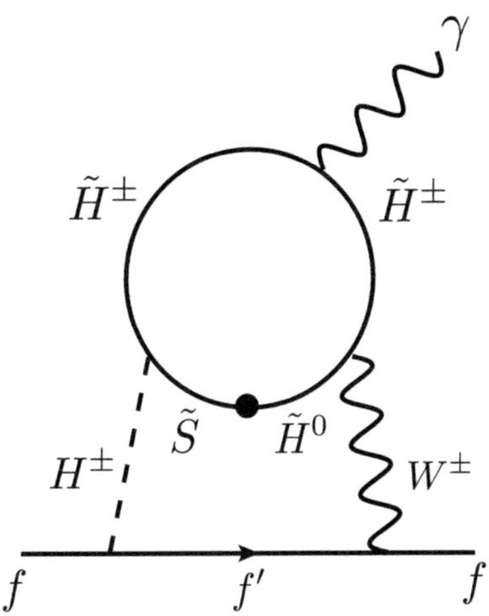

Fig. 5.6 The BAU-related Barr-Zee diagram through the charged scalars and W boson

[3]Here, the Barr-Zee diagram caused by the mixing between the CP-even and CP-odd Higgs is not present since we do not assume the CP violation in the Higgs potential.

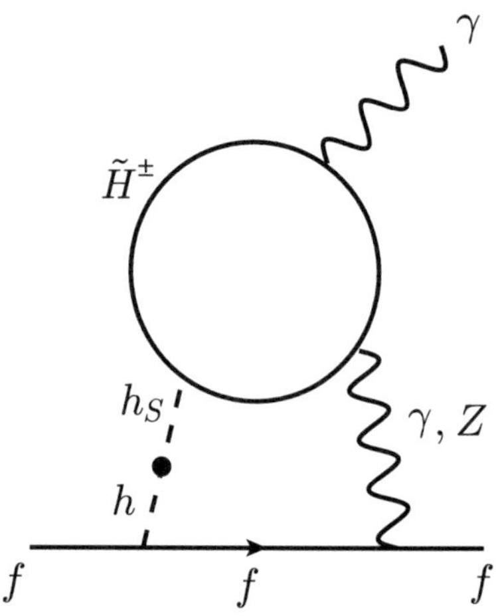

Fig. 5.7 Barr-Zee diagram induced by the BAU-unrelated phase

The CP phases in $g^S_{\bar{\tilde{H}}\tilde{H}S}$ and $g^P_{\bar{\tilde{H}}\tilde{H}S}$ differ from those of $c_L^{\tilde{H}^0\tilde{S}}$ and $c_R^{\tilde{H}^0\tilde{S}}$, and the BAU-unrelated CP phases can induce additional EDMs of $d_f^{H\gamma}$ and d_f^{HZ} as in Fig. 5.7. In total, the EDM is given by

$$\frac{d_f}{e} = \frac{1}{e}\left(d_f^{WW} + d_f^{H^\pm W^\pm} + d_f^{H\gamma} + d_f^{HZ}\right). \tag{5.16}$$

The former two terms are originated from the BAU-related CP phase, and the latter two contributions stem from the BAU-unrelated CP phases. It would be possible to cause the cancellation among them by tuning CP phases in $g^S_{\bar{\tilde{H}}\tilde{H}S}$ and $g^P_{\bar{\tilde{H}}\tilde{H}S}$, and it follows that $d_f/e = 0$. Unfortunately, this situation makes it impossible to verify EWBG in this model by EDMs, even if the successful BAU may be produced. However, it should be noted that the diagrams with the BAU-unrelated CP phase are correlated to the mixing between the $SU(2)$ Higgs and the real singlet. Therefore, we expect that information about the Higgs physics given by upcoming collider experiments can suggest something about the possibility of these contributions.

Although we have shown the formula of d_e^{WW} obtained by mass-insertion method, the full formulas, where the mass matrix of the neutral fermions is diagonalized, are performed in our numerical calculation. After the diagonalization of the mass matrix, the relevant interactions appear

$$\mathcal{L} \ni -\frac{g_2}{\sqrt{2}}\overline{\tilde{\chi}_i^0}\gamma^\mu\left(g^L_{\tilde{\chi}_i^0\tilde{H}^-W^+}P_L + g^R_{\tilde{\chi}_i^0\tilde{H}^-W^+}P_R\right)\tilde{H}^-W^+_\mu$$
$$-\frac{g_2}{\sqrt{2}}\overline{\tilde{H}^-}\gamma^\mu\left(g^L_{\tilde{H}^-\tilde{\chi}_i^0W^-}P_L + g^R_{\tilde{H}^-\tilde{\chi}_i^0W^-}P_R\right)\tilde{\chi}_i^0W^-_\mu, \tag{5.17}$$

where $\tilde{\chi}_i^0$ ($i = 1 \sim 3$) represents the mass eigenstates defined by a complex orthogonal matrix N_χ

$$\frac{1}{2}\left(\tilde{\phi}_1^0\ \tilde{\phi}_2^{0\prime}\ \tilde{S}\right)\left(N_\chi^T N_\chi\right)\mathcal{M}\left(N_\chi N_\chi^T\right)\begin{pmatrix}\tilde{\phi}_1^0\\ \tilde{\phi}_2^{0\prime}\\ \tilde{S}\end{pmatrix} = \left(\tilde{\chi}_1^0\ \tilde{\chi}_2^0\ \tilde{\chi}_3^0\right)\begin{pmatrix} m_{\chi 1} & & \\ & m_{\chi 2} & \\ & & m_{\chi 3}\end{pmatrix}\begin{pmatrix}\tilde{\chi}_1^0\\ \tilde{\chi}_2^0\\ \tilde{\chi}_3^0\end{pmatrix}, \tag{5.18}$$

and it follows that

$$g^L_{\tilde{\chi}_i^0 \tilde{H}^- W^+} = \left(N_\chi^\dagger\right)_{i1}, \qquad g^R_{\tilde{\chi}_i^0 \tilde{H}^- W^+} = \left(N_\chi^\dagger\right)_{2i}, \tag{5.19}$$

$$g^L_{\tilde{H}^- \tilde{\chi}_i^0 W^-} = \left(N_\chi\right)_{1i}, \qquad g^R_{\tilde{H}^- \tilde{\chi}_i^0 W^-} = \left(N_\chi^\dagger\right)_{i2}. \tag{5.20}$$

As shown in Ref.[33–38], these interactions lead to

$$\frac{d_f^{WW}}{e} = \pm\frac{\alpha_{\rm em}^2\,|C|^2}{32\pi^2 s_W^4}\frac{m_f m_{\tilde{\chi}_i^0} m_{\tilde{H}^\pm}}{m_W^4}{\rm Im}\left[g^L_{\tilde{H}^- \tilde{\chi}_i^0 W^-}\left(g^R_{\tilde{H}^- \tilde{\chi}_i^0 W^-}\right)^*\right] f_{WW}(r_{\tilde{\chi}_i^0}, r_+), \tag{5.21}$$

where $r_{\tilde{\chi}_i^0} = m^2_{\tilde{\chi}_i^0}/m_W^2$ and the index of i runs form 1 to 3. The positive (negative) sign is for down-type (up-type) fermions, and the coefficient of C corresponds to the CKM matrix.

As shown in Appendix A.1, the couplings that induce $d_f^{H\gamma}$ are

$$g^S_{\bar{\tilde{H}}\tilde{H}S} = |\lambda|\cos(\phi_\lambda - \phi_{\tilde{H}}), \qquad g^P_{\bar{\tilde{H}}\tilde{H}S} = -|\lambda|\cos(\phi_\lambda - \phi_{\tilde{H}}). \tag{5.22}$$

With these interactions, the Higgs-photon-mediated Barr-Zee diagram is given by

$$\frac{d_f^{H\gamma}}{e} = \frac{\alpha_{\rm em}}{8\pi^3}\frac{Q_f m_f}{m_{\tilde{H}^\pm} v}|\lambda|\sin\phi_{\lambda\tilde{H}}\sin\gamma\cos\gamma\left[g\left(\frac{m^2_{\tilde{H}^\pm}}{m^2_{H_1}}\right) - g\left(\frac{m^2_{\tilde{H}^\pm}}{m^2_{H_2}}\right)\right], \tag{5.23}$$

where $\phi_{\lambda\tilde{H}} = \phi_\lambda - \phi_{\tilde{H}}$, and

$$g(t) = \frac{t}{2}\int_0^1 dx\,\frac{1}{x(1-x)-t}\log\left[\frac{x(1-x)}{t}\right]. \tag{5.24}$$

In our numerical analysis, we drop d_e^{HZ} since it involves accidentally suppressed factor of $(1/2 - \sin^2\theta_W) \simeq 0.02$.

Finally, the $H^{\pm}W^{\mp}$-type diagrams are given by

$$\frac{d_f^{H^{\pm}W^{\pm}}}{e} = \pm \frac{\alpha_{\rm em}}{(4\pi)^3 \sin\theta_W^2} \frac{m_f}{\sqrt{2}v} \frac{1}{m_{H^{\pm}}^2}$$

$$\int_0^1 dx \, \frac{1}{1-x} J\left(r_{WH^{\pm}}, \frac{r_{\tilde{\chi}^{\pm}H^{\pm}}}{1-x} + \frac{r_{\tilde{\chi}_i^0 H^{\pm}}}{x}\right)$$

$$\times \left[\mathrm{Im}\left\{ \left(g^S_{H^+ f_u f_d} + i g^P_{H^+ f_u f_d} \right) G_+^{LR} \right\} m_{\tilde{\chi}_i^0} (1-x)^2 \right.$$

$$+ \mathrm{Im}\left\{ \left(g^S_{H^+ f_u f_d} + i g^P_{H^+ f_u f_d} \right) G_+^{RL} \right\} m_{\tilde{H}^{\pm}} x^2$$

$$\times \mathrm{Im}\left\{ \left(g^S_{H^+ f_u f_d} + i g^P_{H^+ f_u f_d} \right) G_-^{LR} \right\} m_{\tilde{\chi}_i^0} (1-x)$$

$$\left. + \mathrm{Im}\left\{ \left(g^S_{H^+ f_u f_d} + i g^P_{H^+ f_u f_d} \right) G_-^{RL} \right\} m_{\tilde{H}^{\pm}} x \right]. \quad (5.25)$$

where f_u (f_d) represents up-type (down-type) fermion and

$$G_-^{LR} = \left(g^S_{\tilde{\chi}\tilde{H}^{\pm}H^{\pm}} \right)^* \left(g^{L*}_{\tilde{H}^{\pm}\tilde{\chi}W^{\pm}} - g^{R*}_{\tilde{H}^{\pm}\tilde{\chi}W^{\pm}} \right)$$
$$+ i \left(g^P_{\tilde{\chi}\tilde{H}^{\pm}H^{\pm}} \right)^* \left(g^{L*}_{\tilde{H}^{\pm}\tilde{\chi}W^{\pm}} + g^{R*}_{\tilde{H}^{\pm}\tilde{\chi}W^{\pm}} \right), \quad (5.26)$$

$$G_-^{RL} = \left(g^S_{\tilde{\chi}\tilde{H}^{\pm}H^{\pm}} \right)^* \left(g^{R*}_{\tilde{H}^{\pm}\tilde{\chi}W^{\pm}} - g^{L*}_{\tilde{H}^{\pm}\tilde{\chi}W^{\pm}} \right)$$
$$+ i \left(g^P_{\tilde{\chi}\tilde{H}^{\pm}H^{\pm}} \right)^* \left(g^{R*}_{\tilde{H}^{\pm}\tilde{\chi}W^{\pm}} + g^{L*}_{\tilde{H}^{\pm}\tilde{\chi}W^{\pm}} \right), \quad (5.27)$$

$$G_+^{LR} = \left(g^S_{\tilde{\chi}\tilde{H}^{\pm}H^{\pm}} \right)^* \left(g^{L*}_{\tilde{H}^{\pm}\tilde{\chi}W^{\pm}} + g^{R*}_{\tilde{H}^{\pm}\tilde{\chi}W^{\pm}} \right)$$
$$+ i \left(g^P_{\tilde{\chi}\tilde{H}^{\pm}H^{\pm}} \right)^* \left(g^{L*}_{\tilde{H}^{\pm}\tilde{\chi}W^{\pm}} - g^{R*}_{\tilde{H}^{\pm}\tilde{\chi}W^{\pm}} \right), \quad (5.28)$$

$$G_+^{RL} = \left(g^S_{\tilde{\chi}\tilde{H}^{\pm}H^{\pm}} \right)^* \left(g^{R*}_{\tilde{H}^{\pm}\tilde{\chi}W^{\pm}} + g^{L*}_{\tilde{H}^{\pm}\tilde{\chi}W^{\pm}} \right)$$
$$+ i \left(g^P_{\tilde{\chi}\tilde{H}^{\pm}H^{\pm}} \right)^* \left(g^{R*}_{\tilde{H}^{\pm}\tilde{\chi}W^{\pm}} - g^{L*}_{\tilde{H}^{\pm}\tilde{\chi}W^{\pm}} \right), \quad (5.29)$$

with

$$g^S_{\tilde{\chi}\tilde{H}^{\pm}H^{\pm}} = \frac{1}{2} \left(g^L_{\tilde{\chi}\tilde{H}^{\pm}H^{\pm}} + g^R_{\tilde{\chi}\tilde{H}^{\pm}H^{\pm}} \right)$$
$$= \frac{1}{2} \left\{ -c^L_{\tilde{H}^0\tilde{S}} (N_\chi)_{3i} \cos\beta + c^R_{\tilde{H}^0\tilde{S}} (N_\chi)^*_{3i} \sin\beta \right\}, \quad (5.30)$$

$$g^{P}_{\tilde{\chi}\tilde{H}^{\pm}H^{\pm}} = \frac{i}{2}\left(g^{L}_{\tilde{\chi}\tilde{H}^{\pm}H^{\pm}} - g^{R}_{\tilde{\chi}\tilde{H}^{\pm}H^{\pm}}\right)$$

$$= \frac{i}{2}\left\{-c^{L}_{\tilde{H}^{0}\tilde{S}}(N_{\chi})_{3i}\cos\beta - c^{R}_{\tilde{H}^{0}\tilde{S}}(N_{\chi})^{*}_{3i}\sin\beta\right\}, \qquad (5.31)$$

$$g^{L*}_{\tilde{H}^{\pm}\tilde{\chi}W^{\pm}} = (N_{\chi})^{*}_{1i}, \qquad (5.32)$$

$$g^{R*}_{\tilde{H}^{\pm}\tilde{\chi}W^{\pm}} = (N_{\chi})_{2i}. \qquad (5.33)$$

References

1. M. Pospelov, A. Ritz, Ann. Phys. **318**, 119 (2005) [hep-ph/0504231]
2. T. Fukuyama, Int. J. Mod. Phys. A **27**, 1230015 (2012) [arXiv:1201.4252 [hep-ph]]
3. J. Engel, M.J. Ramsey-Musolf, U. van Kolck, Prog. Part. Nucl. Phys. **71**, 21 (2013) [arXiv:1303.2371 [nucl-th]]
4. N. Yamanaka, "Analysis of the Electric Dipole Moment in the R-parity Violating Supersymmetric Standard Model," Springer Theses (2014)
5. 2015 Nuclear Science Advisory Committee Long Range Plan "Reaching for the Horizon", https://science.energy.gov/~/media/np/nsac/pdf/2015LRP/2015_LRPNS_091815.pdf
6. E.P. Shabalin, Sov. J. Nucl. Phys. **28**, 75 (1978) [Yad. Fiz. **28**, 151 (1978)]
7. A. Czarnecki, B. Krause, Phys. Rev. Lett. **78**, 4339 (1997) [hep-ph/9704355]
8. I.B. Khriplovich, Phys. Lett. B **173**, 193 (1986) [Sov. J. Nucl. Phys. **44**, 659 (1986)] [Yad. Fiz. **44**, 1019 (1986)]
9. M.B. Gavela, A. Le Yaouanc, L. Oliver, O. Pene, J.C. Raynal, T.N. Pham, Phys. Lett. B **109**, 215 (1982)
10. I.B. Khriplovich, A.R. Zhitnitsky, Phys. Lett. B **109**, 490 (1982)
11. M.E. Pospelov, I.B. Khriplovich, Sov. J. Nucl. Phys. **53**, 638 (1991) [Yad. Fiz. **53**, 1030 (1991)]
12. J.M. Pendlebury et al., Phys. Rev. D **92**(9), 092003 (2015) [arXiv:1509.04411 [hep-ex]]
13. K. Kumar, Z.T. Lu, M.J. Ramsey-Musolf, arXiv:1312.5416 [hep-ph]
14. B. Graner, Y. Chen, E.G. Lindahl, B.R. Heckel, Phys. Rev. Lett. **116**(16), 161601 (2016); Erratum: [Phys. Rev. Lett. **119**(11), 119901 (2017)] [arXiv:1601.04339 [physics.atom-ph]]
15. R.H. Parker et al., Phys. Rev. Lett. **114**(23), 233002 (2015) [arXiv:1504.07477 [nucl-ex]]
16. J. Baron et al., [ACME Collaboration], Science **343**, 269 (2014) [arXiv:1310.7534 [physics.atom-ph]]
17. J. Baron et al., [ACME Collaboration], New J. Phys. **19**, 073029 (2017)
18. T. Chupp, M. Ramsey-Musolf, Phys. Rev. C **91**(3), 035502 (2015) [arXiv:1407.1064 [hep-ph]]
19. V. Cirigliano, W. Dekens, J. de Vries, E. Mereghetti, Phys. Rev. D **94**(3), 034031 (2016) [arXiv:1605.04311 [hep-ph]]
20. W.B. Cairncross et al., Phys. Rev. Lett. **119**(15), 153001 (2017) [arXiv:1704.07928 [physics.atom-ph]]
21. T. Fleig, M. Jung, JHEP **1807**, 012 (2018) [arXiv:1802.02171 [hep-ph]]
22. K. Fuyuto, M. Ramsey-Musolf, T. Shen, arXiv:1804.01137 [hep-ph]
23. U. Baur, A. Juste, L.H. Orr, D. Rainwater, Phys. Rev. D **71**, 054013 (2005)
24. U. Baur, A. Juste, L.H. Orr, D. Rainwater, Nucl. Phys. Proc. Suppl. **160**, 17 (2006)
25. A.O. Bouzas, F. Larios, Phys. Rev. D **87**(7), 074015 (2013)
26. M. Fael, T. Gehrmann, Phys. Rev. D **88**, 033003 (2013)
27. S. Fayazbakhsh, S.T. Monfared, M. Mohammadi Najafabadi, Phys. Rev. D **92**(1), 014006 (2015)
28. S.M. Etesami, S. Khatibi, M. Mohammadi Najafabadi, Eur. Phys. J. C **76**(10), 533 (2016)

29. A. Cordero-Cid, J.M. Hernandez, G. Tavares-Velasco, J.J. Toscano, J. Phys. G **35**, 025004 (2008)
30. J.F. Kamenik, M. Papucci, A. Weiler, Phys. Rev. D **85**, 071501 (2012); Erratum: [Phys. Rev. D **88**, no. 3, 039903 (2013)]
31. V. Cirigliano, W. Dekens, J. de Vries, E. Mereghetti, Phys. Rev. D **94**(1), 016002 (2016)
32. K. Fuyuto, M. Ramsey-Musolf, Phys. Lett. B **781**, 492 (2018) [arXiv:1706.08548 [hep-ph]].
33. T.H. West, Phys. Rev. D **50**, 7025 (1994).
34. D. Chang, W.F. Chang, W.Y. Keung, Phys. Rev. D **71**, 076006 (2005) [hep-ph/0503055]
35. G.F. Giudice, A. Romanino, Phys. Lett. B **634**, 307 (2006) [hep-ph/0510197]
36. J.R. Ellis, J.S. Lee, A. Pilaftsis, JHEP **0810**, 049 (2008) [arXiv:0808.1819 [hep-ph]]
37. K. Cheung, O.C.W. Kong, J.S. Lee, JHEP **0906**, 020 (2009) [arXiv:0904.4352 [hep-ph]]
38. J. Ellis, J.S. Lee, A. Pilaftsis, JHEP **1010**, 049 (2010) [arXiv:1006.3087 [hep-ph]]
39. K. Fuyuto, J. Hisano, E. Senaha, Phys. Lett. B **755**, 491 (2016) [arXiv:1510.04485 [hep-ph]]
40. P.A.R. Ade et al., [Planck Collaboration], Astron. Astrophys. **571**, A16 (2014) [arXiv:1303.5076 [astro-ph.CO]]

Chapter 6
Results

Abstract So far, we have seen how our model works for the first-order EWPT and produces the baryon number. The Higgs physics has a possibility that the LHC can survey the feasible region for the first-order EWPT. Also, the BAU-related CP phase can yield the EDMs through the two-loop diagrams. In this chapter, taking into account for experimental constraints, we present our numerical results of both EWPT and the BAU.

Keywords Baryon number conservation condition · Higgs coupling measurements · Electric dipole moments

6.1 Baryon Number Conservation Condition

One requirement for the successful EWBG is the baryon number conservation condition, $v_C/T_C > \zeta_{\text{sph}}(T_C)$. Figure 6.1 shows the feasible region that satisfies the condition as a function of the real singlet mass m_{H_2}. The vertical axis indicates the Higgs couplings to the gauge bosons and fermions defined in Eq. (3.33), where it is defined that $\kappa = \kappa_F = \kappa_V$. As κ is deviated from 1, the successful region becomes wide. This behavior is naively understood as follows: a large mixing between H_1 and H_2 causes the deviation of κ from 1, which implies a sizable value of $\mu_{HS} + \lambda_{HS} v_S$. This extension yields the wide region as m_{H_2} is heavy. If we take into account of the value of κ_V in Eq. (3.34), the feasible region implies that $m_{H_2} < 170\,\text{GeV}$.

We also consider the deviation of the Higgs triple coupling from the SM value, which can be measurable at the International Linear Collider [2, 3]. It is defined by

---

This chapter is based on the following two articles: Kaori Fuyuto and Eibun Senaha, Phys. Rev. D **90**, 015015, Copyright ©2014 American Physical Society, Kaori Fuyuto, Junji Hisano and Eibun Senaha, Phys. Lett. B **755**, 491 (2016), Copyright ©2016 The Authors. Published by Elsevier B.V.

© Springer Nature Singapore Pte Ltd. 2018

K. Fuyuto, *Electroweak Baryogenesis and Its Phenomenology*, Springer Theses,
https://doi.org/10.1007/978-981-13-1008-9_6

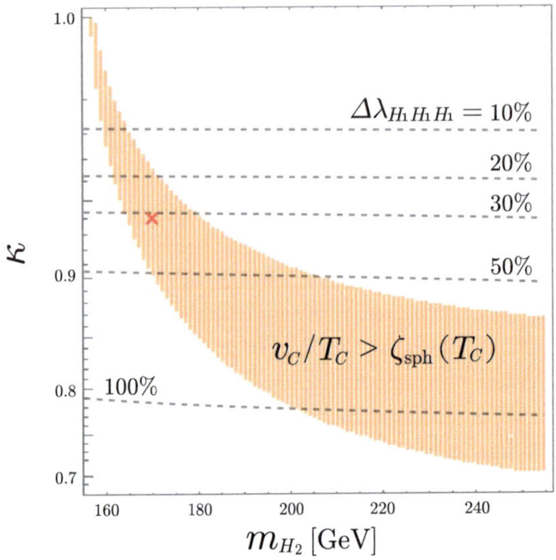

Fig. 6.1 The region where the strong first-order EWPT is satisfied in (m_{H_2}, κ) plane. Here, it is defined that $\kappa = \kappa_F = \kappa_V$. The dashed lines indicate the deviation of the Higgs triple coupling from the SM value, $\Delta\lambda_{H_1H_1H_1} = 10\%,\ 20\%,\ 30\%,\ 50\%$, and 100%, from top to bottom. The red cross mark is a benchmark point for numerical calculations of the BAU. This is studied in a paper, Kaori Fuyuto and Eibun Senaha, Phys. Rev. D **90**, 015015 [1]. (Copyright ©2014 American Physical Society.)

$$\Delta\lambda_{H_1H_1H_1} = \frac{\lambda^{\rm rSM}_{H_1H_1H_1} - \lambda^{\rm SM}_{H_1H_1H_1}}{\lambda^{\rm SM}_{H_1H_1H_1}}. \tag{6.1}$$

The deviation at tree level is expressed by

$$\lambda^{\rm rSM,tree}_{H_1H_1H_1} = 6\left[\lambda_H v c_\gamma^3 + \frac{\mu_{HS}}{2} s_\gamma c_\gamma^2 + \frac{\lambda_{HS}}{2} s_\gamma c_\gamma (v s_\gamma + v_S c_\gamma) + \left(\frac{\mu'_S}{3} + \lambda_S v_S\right) s_\gamma^3\right], \tag{6.2}$$

and it is seen that it becomes nonzero if mixing exists. In Fig. 6.1, the dashed lines represent $\Delta\lambda_{H_1H_1H_1} = 10\%,\ 20\%,\ 30\%,\ 50\%$, and 100%, from top to bottom. Here, the deviation of the Higgs triple coupling is estimated at 1-loop level. As the successful region becomes wider, the deviation of the Higgs triple coupling also becomes larger.

The red cross mark indicates a benchmark point that we use for the estimation of the BAU, and its detail is shown in Table 6.1.

Table 6.1 A benchmark point indicated by a red cross mark in Fig. 6.1

m_{H_2} [GeV]	170
γ	$-20°$
v_S [GeV]	90
μ_{HS} [GeV]	-80
μ'_S [GeV]	-30
λ_H	0.13
λ_S	0.23
λ_{HS}	1.08
κ_V, κ_F	0.94
$\Delta\lambda_{H_1H_1H_1}$ [%]	31.8
v_C/T_C	$206.75/111.76 = 1.85$
$\mathcal{E}(T_C)$	1.80
$\zeta_{\rm sph}(T_C)$	1.18

6.2 Baryon Number with the BAU-Related CP Phase

Before we see our numerical results for the BAU, we comment on the input parameters. As seen in the CP-violating source term, numerical calculations need to extract profiles of the Higgs VEVs, namely, $v(\bar{z})$ and $\beta(\bar{z})$. They can be determined by finding static solutions for bubbles at the nucleation temperature and are often approximated by the kink-type configurations

$$v(\bar{z}) = \frac{v_C}{2}\left[1 - \tanh\left\{\alpha\left(1 - \frac{2\bar{z}}{L_w}\right)\right\}\right], \tag{6.3}$$

$$\beta(\bar{z}) = \beta^{\rm br}(T_C) - \frac{\Delta\beta}{2}\left[1 + \tanh\left\{\alpha\left(1 - \frac{2\bar{z}}{L_w}\right)\right\}\right], \tag{6.4}$$

and the latter leads to

$$\dot{\beta} = \frac{\partial\beta(\bar{z})}{\partial t} = v_w\frac{\partial\beta(\bar{z})}{\partial\bar{z}} = \frac{\alpha}{\cosh^2\left\{\alpha\left(1 - \frac{2\bar{z}}{L_w}\right)\right\}}\left(\frac{v_w\Delta\beta}{L_w}\right). \tag{6.5}$$

In Ref. [4], bubble profiles are estimated in the framework of the MSSM. It is found that $\dot{\beta}$ seems to scale as $1/m_A^2$, where m_A is the mass of the CP-odd Higgs. Although both $v(\bar{z})$ and $\dot{\beta}$ are really important for estimation of the BAU with better accuracy, we take feasible values

$$v(X) = \frac{v_C}{2}, \qquad \dot{\beta} = \frac{v_w\Delta\beta}{L_w}, \tag{6.6}$$

with $\Delta\beta = 0.015$ [4] and $v_w = 0.04$ [5]. These assumptions allow n_B to be independent of L_w.

For thermal widths, we follow those used in [6]

$$\Gamma^t_{\tilde{H}} = 0.0025T, \qquad \Gamma^t_{\tilde{S}} = 0.003T. \tag{6.7}$$

Table 6.2 Input parameters involved in the two Higgs doublet model

$\tan\beta$	1
m_{H_1} [GeV]	125
$m_A = m_H = m_{H^\pm}$ [GeV]	400
$M^2 \equiv m_3^2/(\sin\beta\cos\beta)$ [GeV^2]	1.25×10^5

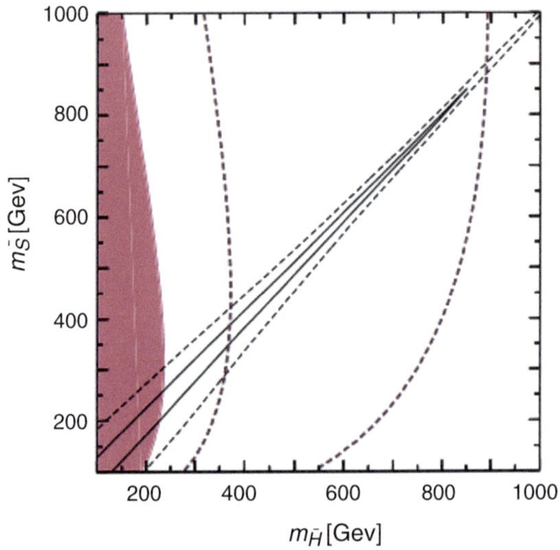

Fig. 6.2 Contour of Y_B/Y_B^{jobs} and $|d_e|$ in the $(m_{\tilde{H}},\ m_{\tilde{S}})$ plane. The black solid (dashed) contour is $Y_B/Y_B^{\text{obs}} = 1$ and 0.1. The pink region corresponds to the current restriction of the electron EDM, and the pink dashed lines are the predictions of $|d_e| = |d_e^{\text{exp}}|/2$ and $|d_e| = 1.0 \times 10^{-29}\ e \cdot \text{cm}$. It is taken that $|c_L^{\tilde{H}^0\tilde{S}}| = |c_R^{\tilde{H}^0\tilde{S}}| = 0.42$ and $\phi = 225°$. This is partly studied in a paper, Kaori Fuyuto, Junji Hisano and Eibun Senaha, Phys. Lett. B **755**, 491 (2016) [11]. (Copyright ©2016 The Authors. Published by Elsevier B.V.)

The weak and strong sphaleron rates are given by [7]

$$\Gamma_B^{(s)} = 6\kappa\alpha_2^5 T, \qquad \Gamma_{\text{ss}} = 16\kappa'\alpha_s^4 T, \tag{6.8}$$

where $\alpha_s = g_s^2/(4\pi)$ with the strong coupling constant g_s, $\kappa = 20$ and $\kappa' = 1$. It is assumed that all heavy scalars in the 2HDM are degenerate, namely, $m_A = m_H = m_{H^\pm}$, where m_H and $m_{H^\pm}$ are the CP-even and charged Higgs masses, respectively. We list the input parameters involved in the 2HDM in Table 6.2.

In what follows, the type-B of Z_2 assignment for the new fermions is employed in estimations of the electron EDM. We firstly consider the case where only the BAU-related CP phases are present, in which the $WW-$ and $HW-$mediated diagrams in Figs. 5.2 and 5.6 contribute to the EDM, namely, $d_f = d_f^{WW} + d_f^{H^\pm W^\pm}$. Figure 6.2 shows the size of Y_B/Y_B^{obs} in the $(m_{\tilde{H}},\ m_{\tilde{S}})$ plane. Here, we take $|c_L^{\tilde{H}^0\tilde{S}}| = |c_R^{\tilde{H}^0\tilde{S}}| = 0.42$ and $\phi = 225°$. The successful region that satisfies $Y_B/Y_B^{\text{obs}} = 1$ is on the black line, and the black dashed line corresponds to $Y_B/Y_B^{\text{obs}} = 0.1$. The pink region is the excluded one by the experimental limit on the electron EDM, $|d_e^{\exp}| < 8.7 \times 10^{-29}\ e \cdot \text{cm}$. The pink dashed lines represent predictions of $|d_e| = |d_e^{\exp}|/2$ and $|d_e| = 1.0 \times 10^{-29}\ e\,\text{cm}$ from left to right. The latter value can be reachable at the upcoming precision measurements [8–10]. While the heavy mass region of $m_{\tilde{S}}$ receives severe constraint from the electron EDM, the heavy $m_{\tilde{H}^0}$ does not encounter such a situation. The difference between them can be understood from their behaviors in the large mass limits as seen in Fig. 5.3. The BAU is sufficiently generated in the region where two masses are degenerate, which stems from the resonant effect in the CP-violating source term. It seems that the sufficient production might be difficult without such an enhancement effect. The black contour can completely be covered by the future sensitivity of the electron EDM even if the estimation of the BAU is underestimated or overestimated by a factor of 10 owing to the absence of bubble profiles.

6.3 Presence of the BAU-Unrelated CP Phase

When the BAU-unrelated CP phase is introduced, the relation between the BAU and EDMs is not straightforward. In this case, the fermion EDM is described by a combination of $d_f = d_f^{WW} + d_f^{H^\pm W^\pm} + d_f^{H\gamma} + d_f^{HZ}$. In order to understand how the EDMs work for the verification of the scenario, we see the dependence of d_e on the BAU-unrelated parameters. Figure 6.3 presents excluded regions by the electron EDM in $|\lambda|$ and $\phi_{\lambda\tilde{H}}$ plane, where we utilize one benchmark point in Fig. 6.2, $m_{\tilde{H}} = 300\,\text{GeV}$ and $m_{\tilde{S}} = 277\,\text{GeV}$, which can achieve $Y_B/Y_B^{\text{obs}} = 1$. In the left figure, the pink region is excluded by the current limit, and the pink dashed line implies an accidental cancelation between the BAU-related and BAU-unrelated CP phases.

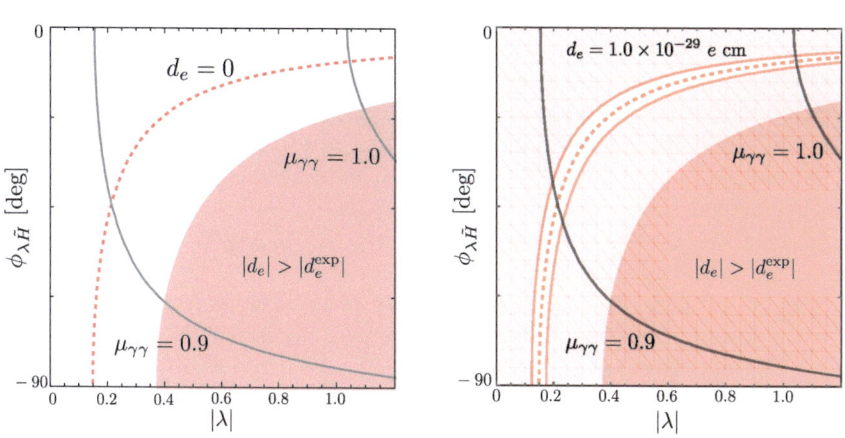

Fig. 6.3 (Left) current excluded region by the current electron EDM limit in $(|\lambda|, \phi_{\lambda\tilde{H}})$ plane at one successful benchmark point with $m_{\tilde{H}^0} = 300\,\text{GeV}$ and $m_{\tilde{S}} = 277\,\text{GeV}$. The pink dashed line represents that $|d_e| = 0$, and the gray lines show $\mu_{\gamma\gamma} = 1.0$ and 0.9. (Right) prospective future excluded region by the limit of $d_e = 1.0 \times 10^{-29}\ e\,\text{cm}$. These are partly studied in a paper, Kaori Fuyuto, Junji Hisano, and Eibun Senaha, Phys. Lett. B **755**, 491 (2016) [11]. (Copyright ©2016 The Authors. Published by Elsevier B.V.)

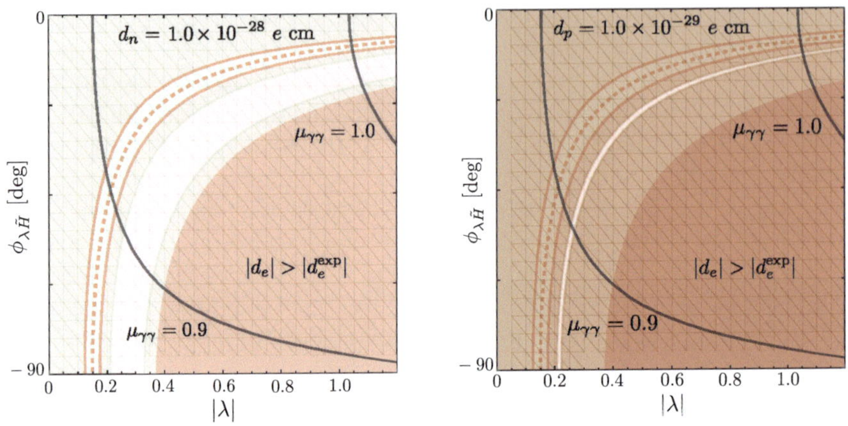

Fig. 6.4 Prospects of the excluded regions by the neutron (left) and proton (right) EDMs. The green and brown regions correspond to the future sensitivities of $d_n = 1.0 \times 10^{-28}$ $e\,$cm and $d_p = 1.0 \times 10^{-29}$ $e\,$cm, respectively

The right figure reflects the future sensitivity with $d_e = 1.0 \times 10^{-29}$ $e\,$cm, where the light pink region can be excluded if the measurement reaches the value. Although the electron EDM expects to cover most of the parameter region, the cancelation region still remains.

One promising approach to examine the region with $d_e = 0$ is the Higgs phenomenology. The $d_e^{H\gamma}$ and d_e^{HZ} contributions disappear if the mixing between neutral scalars is zero, namely, $\gamma = 0$. Moreover, as seen in the first loop in Fig. 5.7, the two-loop diagram is related to the Higgs decay to two gammas. Therefore, the possibility of the contributions from $d_e^{H\gamma}$ and d_e^{HZ} can be examined by both the measurements of the Higgs coupling and the Higgs signal strength. The estimation for the signal strength is discussed in Appendix A.2. The current experimental value which combines Run 1 data from ATLAS and CMS is reported as $\mu_{\gamma\gamma} = 1.14^{+0.19}_{-0.18}$ [12]. In Fig. 6.3, the gray lines correspond to $\mu_{\gamma\gamma} = 1.0$ and 0.9 from top to bottom. If we take into account for the 1σ error of $\mu_{\gamma\gamma}$, only right upper region is allowed. Future colliders such as the high-luminosity LHC [13, 14], ILC [2, 3], and TLEP [15] aim to improve the sensitivity of $\mu_{\gamma\gamma}$ up to $\mathcal{O}(5)\%$ and the Higgs coupling to the gauge bosons up to $\mathcal{O}(0.1)\%$, respectively.

Finally, we consider the future sensitivity of the neutron and proton EDMs. They consist of the up- and down-quark EDMs and chromo EDMs, and our current setup induces only the quark EDMs. Employing expressions in [16], we present the prospective limits on the neutron and proton EDMs in Fig. 6.4. The green region in the left figure is excluded by the neutron EDM with $d_n = 1.0 \times 10^{-28}$ $e\,$cm, while the right figure shows the future sensitivity of $d_p = 1.0 \times 10^{-29}$ $e\,$cm in the brown region. If their sensitivities are achieved, the cancelation region of d_e can be completely covered. Thus, it is understood that the combinations of the precision measurements play powerful role in probing the scenario.

References

1. K. Fuyuto, E. Senaha, Phys. Rev. D **90**(1), 015015 (2014) [arXiv:1406.0433 [hep-ph]]
2. H. Baer, T. Barklow, K. Fujii, Y. Gao, A. Hoang, S. Kanemura, J. List, H.E. Logan et al., arXiv:1306.6352 [hep-ph]
3. T. Barklow, J. Brau, K. Fujii, J. Gao, J. List, N. Walker, K. Yokoya, arXiv:1506.07830 [hep-ex]
4. J.M. Moreno, M. Quiros, M. Seco, Nucl. Phys. B **526**, 489 (1998) [hep-ph/9801272]
5. G.D. Moore, T. Prokopec, Phys. Rev. D **52**, 7182 (1995) [hep-ph/9506475]
6. D.J.H. Chung, B. Garbrecht, M.J. Ramsey-Musolf, S. Tulin, JHEP **0912**, 067 (2009) [arXiv:0908.2187 [hep-ph]]
7. E. Senaha, Phys. Rev. D **88**(5), 055014 (2013) [arXiv:1308.3389 [hep-ph]]
8. Y. Sakemi, K. Harada, T. Hayamizu, M. Itoh, H. Kawamura, S. Liu, H.S. Nataraj, A. Oikawa et al., J. Phys. Conf. Ser. **302**, 012051 (2011)
9. D.M. Kara, I.J. Smallman, J.J. Hudson, B.E. Sauer, M.R. Tarbutt, E.A. Hinds, New J. Phys. **14**, 103051 (2012)
10. D. Kawall, J. Phys. Conf. Ser. **295**, 012031 (2011)
11. K. Fuyuto, J. Hisano, E. Senaha, Phys. Lett. B **755**, 491 (2016) [arXiv:1510.04485 [hep-ph]]
12. G. Aad et al., [ATLAS and CMS Collaborations], JHEP **1608**, 045 (2016) [arXiv:1606.02266 [hep-ex]]
13. [ATLAS Collaboration], arXiv:1307.7292 [hep-ex]
14. [CMS Collaboration], arXiv:1307.7135
15. M. Bicer et al., [TLEP Design Study Working Group Collaboration], JHEP **1401**, 164 (2014)
16. J. Hisano, D. Kobayashi, W. Kuramoto, T. Kuwahara, JHEP **1511**, 085 (2015) [arXiv:1507.05836 [hep-ph]]

Chapter 7
Conclusion

In this thesis, we have studied the possibility of electroweak baryogenesis in a model where an extra Higgs doublet, a real singlet, and new EW-interacting fermions are introduced. The real singlet gives important contributions to the Higgs potential for inducing the first-order electroweak phase transition, whereas the new fermions supply relevant CP phases to producing baryon asymmetry. In this setup, the Higgs phenomenology and precision measurements of electric dipole moments are expected to be able to examine the possibility.

Achieving the first-order phase transition needs nonzero mixing terms between the real singlet and the Higgs scalars. The presence of the mixing terms at tree level affects the Higgs boson couplings, and they can be deviated from those of the SM values. We found that one moderate benchmark point in the current model shows that the deviation of the Higgs couplings to gauge bosons and fermions becomes 6%. Moreover, that of the Higgs triple coupling amounts to about 30%. These deviations can be probed in future collider experiments.

The dependence on the CP phase differs between the electric dipole moments and baryon number. The former is simply proportional to the sine function of the CP phase; however, the dependence in the baryon asymmetry appears not only in the CP-violating terms but also in the CP-conserving terms. Thus, although the maximum values of the phase leads to those of the electric dipole moments of fermions, the baryon number cannot be produced maximally. Moreover, if another phase is not involved in the production of the asymmetry but associated with the low-energy observables, the phase renders the situation more complicated.

Our model owes two CP phases: one is the BAU-related, and the other is the BAU-unrelated one. The BAU-related phase arises from interactions between the new fermions and the space-dependent VEVs. Since our estimations of the baryon numbers employ VEV-insertion method, only regions where the two fermions are degenerate can produce enough amount of the asymmetry to explain the current observed value. As long as only the BAU-related CP phase is present, the successful

© Springer Nature Singapore Pte Ltd. 2018

K. Fuyuto, *Electroweak Baryogenesis and Its Phenomenology*, Springer Theses,
https://doi.org/10.1007/978-981-13-1008-9_7

region can be completely probed by the precision measurement of the electron EDM with proposed future sensitivity, $d_e = 1.0 \times 10^{-29}\ e \cdot \text{cm}$.

The interactions between the new fermions and singlet scalar yield the BAU-unrelated CP phase, which contributes to the fermion EDMs through different two-loop diagrams. Once the phase comes into action, the relationship between the BAU and the fermion EDMs breaks down. The fermion EDMs are induced by a combination of four different Barr-Zee diagrams, and the combination can accidentally cancel out implying $d_e = 0$. It indicates that the electron EDM is no longer be able to verify the scenario even if the successful BAU is realized.

One helpful phenomenology in verifying the cancelation region is the Higgs physics. The BAU-unrelated contributions to the electron EDM stem from the mixing between the singlet and Higgs; therefore they can be investigated by the precision measurements of the Higgs couplings as well as the phase transition. Moreover, the BAU-unrelated two-loop diagrams affect the signal strength of the Higgs decay to two gammas. Therefore, we expect that the Higgs physics helps to probe the scenario even if the BAU-unrelated CP phase exists. The other powerful approaches are precise measurements of neutron and proton EDMs, which are composed of quark EDMs. It turns out that, if they reach $d_n = 1.0 \times 10^{-28}\ e$ cm and $d_p = 1.0 \times 10^{-29}\ e$ cm, the cancelation region can be covered.

The Higgs phenomenology in collider experiments and precision measurements of the EDMs are necessary for probing our scenario. We expect that the possibility in the model would be completely examined in the near future.

Appendix

A.1 Four-Component Notation

Four component fields are given by

$$\tilde{S} = \begin{pmatrix} \tilde{S}^0 \\ \bar{\tilde{S}}^0 \end{pmatrix}, \quad \tilde{H}^0 = \begin{pmatrix} \tilde{\phi}_2^0 \\ \bar{\tilde{\phi}}_1^0 \end{pmatrix}, \quad \tilde{H}^+ = \begin{pmatrix} -\tilde{\phi}_2^+ \\ \bar{\tilde{\phi}}_1^+ \end{pmatrix}, \quad \tilde{H}^- = \begin{pmatrix} \tilde{\phi}_1^- \\ -\bar{\tilde{\phi}}_2^- \end{pmatrix}, \tag{1}$$

$$\overline{\tilde{S}} = \begin{pmatrix} \tilde{S}^0 \ \bar{\tilde{S}}^0 \end{pmatrix}, \quad \overline{\tilde{H}^0} = \begin{pmatrix} \tilde{\phi}_1^0 \ \bar{\tilde{\phi}}_2^0 \end{pmatrix},$$

$$\overline{\tilde{H}^+} = (\tilde{H}^-)^\dagger \gamma^0 = \begin{pmatrix} \bar{\tilde{\phi}}_1^+ \ -\tilde{\phi}_2^+ \end{pmatrix} \begin{pmatrix} 0 & \mathbf{1} \\ \mathbf{1} & 0 \end{pmatrix} = \begin{pmatrix} -\tilde{\phi}_2^+ \ \bar{\tilde{\phi}}_1^+ \end{pmatrix},$$

$$\overline{\tilde{H}^-} = (\tilde{H}^+)^\dagger \gamma^0 = \begin{pmatrix} -\bar{\tilde{\phi}}_2^- \ \tilde{\phi}_1^- \end{pmatrix} \begin{pmatrix} 0 & \mathbf{1} \\ \mathbf{1} & 0 \end{pmatrix} = \begin{pmatrix} \tilde{\phi}_1^- \ -\bar{\tilde{\phi}}_2^- \end{pmatrix}. \tag{2}$$

We redefine parameters and fields in the potential so that mass of each field becomes real. For types A and B,

$$\mu_{\tilde{S}} = |\mu_{\tilde{S}}| e^{i\phi_{\tilde{S}}} \equiv m_{\tilde{S}} e^{i\phi_{\tilde{S}}}, \tag{3}$$

$$\tilde{\Phi}_2 = e^{-i\phi_{\tilde{H}}} \tilde{\Phi}_2', \tag{4}$$

$$\tilde{S}^0 = e^{-i\phi_{\tilde{S}}/2} \tilde{S}^{0\prime}, \tag{5}$$

$$\lambda = |\lambda| e^{i\phi_\lambda}, \tag{6}$$

$$\mu + \lambda v_S = |\mu + \lambda v_S| e^{i\,\mathrm{Arg}(\mu + \lambda v_S)} \equiv m_{\tilde{H}} e^{i\phi_{\tilde{H}}}. \tag{7}$$

In what follows, primes in $\tilde{\Phi}'$ and $\tilde{S}'$ are dropped. With these definitions, the Lagrangian for type A becomes

© Springer Nature Singapore Pte Ltd. 2018

K. Fuyuto, *Electroweak Baryogenesis and Its Phenomenology*, Springer Theses,
https://doi.org/10.1007/978-981-13-1008-9

$$
\begin{aligned}
\mathcal{L}_{\text{fer}}^{\text{A}} = {} & \overline{\tilde{H}^{+}}(i\gamma\!\!\!/ - m_{\tilde{H}})\tilde{H}^{+} + \overline{\tilde{H}^{0}}(i\gamma\!\!\!/ - m_{\tilde{H}})\tilde{H}^{0} + \frac{1}{2}\overline{\tilde{S}}(i\gamma\!\!\!/ - m_{\tilde{S}})\tilde{S} \\
& + \left\{ \overline{\tilde{H}^{0}}\left(c_L^{\tilde{H}^0\tilde{S}} P_L \phi_1^0 + c_R^{\tilde{H}^0\tilde{S}} P_R \phi_2^0\right)\tilde{S} \right. \\
& \left. + \overline{\tilde{H}^{+}}\left(c_L^{\tilde{H}^\pm\tilde{S}} P_L \phi_1^+ + c_R^{\tilde{H}^\pm\tilde{S}} P_R \phi_2^+\right)\tilde{S} + (\text{h.c}) \right\} \\
& - \overline{\tilde{H}^{+}}\left(g^S_{\bar{\tilde{H}}\tilde{H}S} + i\gamma_5 g^P_{\bar{\tilde{H}}\tilde{H}S}\right)\tilde{H}^{+} h_S - \overline{\tilde{H}^{0}}\left(g^S_{\bar{\tilde{H}}\tilde{H}S} + i\gamma_5 g^P_{\bar{\tilde{H}}\tilde{H}S}\right)\tilde{H}^{0} h_S, \qquad (8)
\end{aligned}
$$

where

$$c_L^{\tilde{H}^\pm\tilde{S}} = c_{11} e^{-i\phi_{\tilde{S}}/2}, \qquad c_L^{\tilde{H}^0\tilde{S}} = -c_{11} e^{-i\phi_{\tilde{S}}/2}, \qquad (9)$$

$$c_R^{\tilde{H}^\pm\tilde{S}} = -c_{22}^{*} e^{i(\phi_{\tilde{H}} + \phi_{\tilde{S}}/2)}, \qquad c_R^{\tilde{H}^0\tilde{S}} = c_{22}^{*} e^{i(\phi_{\tilde{H}} + \phi_{\tilde{S}}/2)}, \qquad (10)$$

$$g^S_{\bar{\tilde{H}}\tilde{H}S} = |\lambda| \cos(\phi_\lambda - \phi_{\tilde{H}}), \qquad g^P_{\bar{\tilde{H}}\tilde{H}S} = -|\lambda| \sin(\phi_\lambda - \phi_{\tilde{H}}). \qquad (11)$$

For type B,

$$
\begin{aligned}
\mathcal{L}_{\text{fer}}^{\text{B}} = {} & \overline{\tilde{H}^{+}}(i\gamma\!\!\!/ - m_{\tilde{H}})\tilde{H}^{+} + \overline{\tilde{H}^{0}}(i\gamma\!\!\!/ - m_{\tilde{H}})\tilde{H}^{0} + \frac{1}{2}\overline{\tilde{S}}(i\gamma\!\!\!/ - m_{\tilde{S}})\tilde{S} \\
& + \left\{ \overline{\tilde{H}^{0}}\left(c_L^{\tilde{H}^0\tilde{S}} P_L \phi_2^0 + c_R^{\tilde{H}^0\tilde{S}} P_R \phi_1^0\right)\tilde{S} \right. \\
& \left. + \overline{\tilde{H}^{+}}\left(c_L^{\tilde{H}^\pm\tilde{S}} P_L \phi_2^+ + c_R^{\tilde{H}^\pm\tilde{S}} P_R \phi_1^+\right)\tilde{S} + (\text{h.c}) \right\} \\
& - \overline{\tilde{H}^{+}}\left(g^S_{\bar{\tilde{H}}\tilde{H}S} + i\gamma_5 g^P_{\bar{\tilde{H}}\tilde{H}S}\right)\tilde{H}^{+} h_S - \overline{\tilde{H}^{0}}\left(g^S_{\bar{\tilde{H}}\tilde{H}S} + i\gamma_5 g^P_{\bar{\tilde{H}}\tilde{H}S}\right)\tilde{H}^{0} h_S, \qquad (12)
\end{aligned}
$$

where

$$c_L^{\tilde{H}^\pm\tilde{S}} = c_{12} e^{-i\phi_{\tilde{S}}/2}, \qquad c_L^{\tilde{H}^0\tilde{S}} = -c_{12} e^{-i\phi_{\tilde{S}}/2}, \qquad (13)$$

$$c_R^{\tilde{H}^\pm\tilde{S}} = -c_{21}^{*} e^{i(\phi_{\tilde{H}} + \phi_{\tilde{S}}/2)}, \qquad c_R^{\tilde{H}^0\tilde{S}} = c_{21}^{*} e^{i(\phi_{\tilde{H}} + \phi_{\tilde{S}}/2)}, \qquad (14)$$

$$g^S_{\bar{\tilde{H}}\tilde{H}S} = |\lambda| \cos(\phi_\lambda - \phi_{\tilde{H}}), \qquad g^P_{\bar{\tilde{H}}\tilde{H}S} = -|\lambda| \sin(\phi_\lambda - \phi_{\tilde{H}}). \qquad (15)$$

On the other hand, for types C and D, we redefine as follows:

$$\mu = |\mu| e^{i\phi_\mu} \equiv m_{\tilde{H}} e^{i\phi_{\tilde{H}}}, \qquad (16)$$

$$\tilde{\Phi}_2 = e^{-i\phi_{\tilde{H}}} \tilde{\Phi}_2', \qquad (17)$$

$$\tilde{S}^0 = e^{-i\phi_{\tilde{S}}/2} \tilde{S}^{0\prime}, \qquad (18)$$

$$\kappa = |\kappa| e^{i\phi_\kappa}, \qquad (19)$$

$$\mu_{\tilde{S}} + \kappa v_S = |\mu_{\tilde{S}} + \kappa v_S| e^{i\,\text{Arg}(\mu_{\tilde{S}} + \kappa v_S)} \equiv m_{\tilde{S}} e^{i\phi_{\tilde{S}}}. \qquad (20)$$

For type C,

$$
\begin{aligned}
\mathcal{L}_{\text{fer}}^{\text{C}} = {} & \overline{\tilde{H}^+}(i\not{\partial} - m_{\tilde{H}})\tilde{H}^+ + \overline{\tilde{H}^0}(i\not{\partial} - m_{\tilde{H}})\tilde{H}^0 + \frac{1}{2}\overline{\tilde{S}}(i\not{\partial} - m_{\tilde{S}})\tilde{S} \\
& + \Big\{ \overline{\tilde{H}^0}\left(c_L^{\tilde{H}^0\tilde{S}} P_L \phi_1^0 + c_R^{\tilde{H}^0\tilde{S}} P_R \phi_2^0\right)\tilde{S} \\
& + \overline{\tilde{H}^+}\left(c_L^{\tilde{H}^\pm\tilde{S}} P_L \phi_1^+ + c_R^{\tilde{H}^\pm\tilde{S}} P_R \phi_2^+\right)\tilde{S} + (\text{h.c})\Big\} \\
& - \frac{1}{2}\overline{\tilde{S}}\left(g^S_{\bar{\tilde{S}}\tilde{S}S} + i\gamma_5 g^P_{\bar{\tilde{S}}\tilde{S}S}\right)\tilde{S}h_S,
\end{aligned}
\tag{21}
$$

where

$$c_L^{\tilde{H}^\pm\tilde{S}} = c_{11}e^{-i\phi_{\tilde{S}}/2}, \qquad c_L^{\tilde{H}^0\tilde{S}} = -c_{11}e^{-i\phi_{\tilde{S}}/2}, \tag{22}$$

$$c_R^{\tilde{H}^\pm\tilde{S}} = -c_{22}^* e^{i(\phi_{\tilde{H}}+\phi_{\tilde{S}}/2)}, \qquad c_R^{\tilde{H}^0\tilde{S}} = c_{22}^* e^{i(\phi_{\tilde{H}}+\phi_{\tilde{S}}/2)}, \tag{23}$$

$$g^S_{\bar{\tilde{S}}\tilde{S}S} = |\kappa|\cos(\phi_\kappa - \phi_{\tilde{S}}), \qquad g^P_{\bar{\tilde{S}}\tilde{S}S} = -|\kappa|\sin(\phi_\kappa - \phi_{\tilde{S}}). \tag{24}$$

Finally, for type D,

$$
\begin{aligned}
\mathcal{L}_{\text{fer}}^{\text{D}} = {} & \overline{\tilde{H}^+}(i\not{\partial} - m_{\tilde{H}})\tilde{H}^+ + \overline{\tilde{H}^0}(i\not{\partial} - m_{\tilde{H}})\tilde{H}^0 + \frac{1}{2}\overline{\tilde{S}}(i\not{\partial} - m_{\tilde{S}})\tilde{S} \\
& + \Big\{ \overline{\tilde{H}^0}\left(c_L^{\tilde{H}^0\tilde{S}} P_L \phi_2^0 + c_R^{\tilde{H}^0\tilde{S}} P_R \phi_1^0\right)\tilde{S} \\
& + \overline{\tilde{H}^+}\left(c_L^{\tilde{H}^\pm\tilde{S}} P_L \phi_2^+ + c_R^{\tilde{H}^\pm\tilde{S}} P_R \phi_1^+\right)\tilde{S} + (\text{h.c})\Big\} \\
& - \frac{1}{2}\overline{\tilde{S}}\left(g^S_{\bar{\tilde{S}}\tilde{S}S} + i\gamma_5 g^P_{\bar{\tilde{S}}\tilde{S}S}\right)\tilde{S}h_S,
\end{aligned}
\tag{25}
$$

where

$$c_L^{\tilde{H}^\pm\tilde{S}} = c_{12}e^{-i\phi_{\tilde{S}}/2}, \qquad c_L^{\tilde{H}^0\tilde{S}} = -c_{12}e^{-i\phi_{\tilde{S}}/2}, \tag{26}$$

$$c_R^{\tilde{H}^\pm\tilde{S}} = -c_{21}^* e^{i(\phi_{\tilde{H}}+\phi_{\tilde{S}}/2)}, \qquad c_R^{\tilde{H}^0\tilde{S}} = c_{21}^* e^{i(\phi_{\tilde{H}}+\phi_{\tilde{S}}/2)}, \tag{27}$$

$$g^S_{\bar{\tilde{S}}\tilde{S}S} = |\kappa|\cos(\phi_\kappa - \phi_{\tilde{S}}), \qquad g^P_{\bar{\tilde{S}}\tilde{S}S} = -|\kappa|\sin(\phi_\kappa - \phi_{\tilde{S}}). \tag{28}$$

A.2 Signal Strength of the Higgs Decay to Two Photons

Utilizing the discussion in Ref. [1], we estimate the signal strength of the Higgs decay to two gammas as follows:

$$\mu_{\gamma\gamma} = \frac{\sigma(pp \to H_1) \times \mathrm{BR}(H_1 \to 2\gamma)}{\sigma^{\mathrm{SM}}(pp \to H_1) \times \mathrm{BR}^{\mathrm{SM}}(H_1 \to 2\gamma)}$$

$$\simeq \frac{c_\gamma^2 \sigma^{\mathrm{SM}}(pp \to H_1) \times \mathrm{BR}(H_1 \to 2\gamma)}{\sigma^{SM}(pp \to H_1) \times \mathrm{BR}^{\mathrm{SM}}(H_1 \to 2\gamma)}$$

$$= \frac{\Gamma(H_1 \to 2\gamma)}{\Gamma^{\mathrm{SM}}(H_1 \to 2\gamma)} \times \frac{c_\gamma^2 \Gamma_{\mathrm{total}}^{\mathrm{SM}}}{\Gamma_{\mathrm{total}}}, \tag{29}$$

where we assume $\Gamma_{\mathrm{total}} \simeq c_\gamma^2 \Gamma_{\mathrm{total}}^{\mathrm{SM}}$. The decay rates are obtained by

$$\Gamma^{\mathrm{SM}}(H_1 \to 2\gamma) = \frac{\alpha_{\mathrm{em}}^2 m_{H_1}^3}{256\pi^3 v^2} |\mathcal{A}_{\mathrm{SM}}|^2, \tag{30}$$

$$\Gamma(H_1 \to 2\gamma) = \frac{\alpha_{\mathrm{em}}^2 m_{H_1}^3}{256\pi^3 v^2} \left(\left| \mathcal{A}_{\tilde{H}^\pm}^S \right|^2 + |\mathcal{A}_{\tilde{H}^\pm}^P|^2 \right), \tag{31}$$

where $\mathcal{A}_{\mathrm{SM}} = -6.49$ [2] and

$$\mathcal{A}_{\tilde{H}^\pm}^S = \frac{v g_{H_1}^S}{m_{\tilde{H}^\pm}} 2\tau_{\tilde{H}^\pm} \left\{ 1 + (1 - \tau_{\tilde{H}^\pm}) f(\tau_{\tilde{H}^\pm}) \right\}, \tag{32}$$

$$\mathcal{A}_{\tilde{H}^\pm}^P = \frac{v g_{H_1}^P}{m_{\tilde{H}^\pm}} 2\tau_{\tilde{H}^\pm} f(\tau_{\tilde{H}^\pm}), \tag{33}$$

with

$$g_{H_1}^S = |\lambda| \cos\phi_{\lambda\tilde{H}} s_\gamma, \qquad g_{H_1}^P = -|\lambda| \sin\phi_{\lambda\tilde{H}} s_\gamma. \tag{34}$$

The function of $f(\tau_{\tilde{H}^\pm})$ is given by

$$f(\tau_{\tilde{H}^\pm}) = -\frac{1}{2} \int_0^1 \frac{dy}{y} \log\left[\frac{m_{H_1}^2 y(y - 1 + m_{\tilde{H}^\pm}^2)}{m_{\tilde{H}^\pm}^2} \right], \tag{35}$$

with $\tau_{\tilde{H}^\pm} = 4m_{\tilde{H}^\pm}^2 / m_{H_1}^2$. Therefore, the decay rate of $H_1 \to 2\gamma$ is expressed by

$$\Gamma(H_1 \to 2\gamma) = \frac{\alpha_{\mathrm{em}}^2 m_{H_1}^3}{256\pi^3 v^2} \left| c_\gamma \mathcal{A}_{\mathrm{SM}} + \mathcal{A}_{\tilde{H}^\pm}^S + \mathcal{A}_{\tilde{H}^\pm}^P \right|^2$$

$$= \frac{\alpha_{\mathrm{em}}^2 m_{H_1}^3}{256\pi^3 v^2} \left\{ |c_\gamma \mathcal{A}_{\mathrm{SM}} + \mathcal{A}_{\tilde{H}^\pm}^S|^2 + |\mathcal{A}_{\tilde{H}^\pm}^P|^2 \right\}, \tag{36}$$

which leads to

$$\mu_{\gamma\gamma} \simeq \frac{\Gamma(H_1 \to 2\gamma)}{\Gamma^{\mathrm{SM}}(H_1 \to 2\gamma)} = \left| c_\gamma + \frac{\mathcal{A}^S_{\tilde{H}^\pm}}{\mathcal{A}^{\mathrm{SM}}} \right|^2 + \left| \frac{\mathcal{A}^P_{\tilde{H}^\pm}}{\mathcal{A}^{\mathrm{SM}}} \right|^2 . \tag{37}$$

References

1. C.W. Chiang, E. Senaha, Phys. Lett. B **750**, 147 (2015) [arXiv:1508.02891 [hep-ph]]
2. D. McKeen, M. Pospelov, A. Ritz, Phys. Rev. D **86**, 113004 (2012) [arXiv:1208.4597 [hep-ph]]

Curriculum Vitae

Kaori Fuyuto

Contact Details

Address:	Amherst Center for Fundamental Interactions, Department of Physics, University of Massachusetts Amherst Amherst, MA 01003, US
E-mail:	kfuyuto@umass.edu
Home page:	http://kfuyuto.wixsite.com/kfhp

Education and Employment

Sept. 2016 – now	Post-Doctoral Research Associate, Amherst Center for Fundamental Interactions, University of Massachusetts.
April. 2016 – Aug. 2016	JSPS Research Fellowships, Saga University.
Mar. 2016	Doctor of Science (Physics), Nagoya University.
Mar. 2013	Master of Science (Physics), Nagoya University.
Mar. 2011	Bachelor of Arts and Science (Physics), Aichi University of Education.

Research Interests

CP violation, Flavor Physics, and Higgs Physics

Awards and Honors

- Best Poster Award: *"Electroweak phase transition in the singlet-extended SM"* Higgs as a Probe of New Physics 2015, Toyama, Japan, February 15, 2015
- 9th Particle Physics Medal: Young Scientist Award in Theoretical Particle Physics, [*Phys. Rev. D* **90**, *no. 1, 015015 (2014)*] Group on Theory of Elementary Particles, Japan, March 23, 2015

© Springer Nature Singapore Pte Ltd. 2018

K. Fuyuto, *Electroweak Baryogenesis and Its Phenomenology*, Springer Theses,
https://doi.org/10.1007/978-981-13-1008-9

MIX
Papier aus verantwortungsvollen Quellen
Paper from responsible sources
FSC® C105338

If you have any concerns about our products,
you can contact us on
ProductSafety@springernature.com

In case Publisher is established outside the EU,
the EU authorized representative is:
Springer Nature Customer Service Center GmbH
Europaplatz 3, 69115 Heidelberg, Germany

Printed by Libri Plureos GmbH
in Hamburg, Germany